ALGEBRA & INDICES

A Simple Approach Review and Self-Teaching Practice workbook on Algebra and Indices with different Worked Problems.

I0625435

Adegboye Samuel

TABLE OF CONTENTS

Chapter one

LAW OF INDICES

An *index number* is a number which is raised to a power. A simple way to understand this is shown below.

$$"axaxaxa = a^4"$$

From the example above, "a" is referred to as **base,** while the number **"4"** is referred to as **power, index, or exponent.**

Note:

For a better understanding of some solutions to some problems, we have to consider the fundamental laws of indices, which will be analysed below.

The following laws will help your understanding to solve any question in indices.

- **Law 1:** $a^m \times a^n = a^{m+n}$

This may be referred to **as the law of power addition of indices.** You can also have; $a^m \times a^n \times a^o = a^{m+n+o}$ (irrespective of the numbers of indices entity you are multiplying above).

For example

 Given that;

  x = a^2 and y= a^3, find x × y.

Solution

$$x \times y = a^2 \times a^3 = a^{2+3} = a^5$$

This can equal be;

$$x \times y = a^2 \times a^3$$
$$= (a \times a) \times (a \times a \times a)$$
$$= a \times a \times a \times a \times a = a^5$$

But using the law of indices will make the solution easier and faster. For rule 1 to be valid, the bases of both entities must be the same.

- **Law 2:** $a^m \div a^n = a^{m-n}$

For example

Given that;

x = a^5 and y= a^3 , find x ÷ y.

Solution

$$x \div y = a^5 \div a^3 = a^{5-3} = a^2$$

This can equally be;

$$x \div y = a^5 \div a^3 = \frac{a \times a \times a \times a \times a}{a \times a \times a} = a^2$$

- **Law 3:** $(a^x)^y = a^{x \times y} = a^{xy}$

For example:

$$(a^4)^3 = a^{4 \times 3} = a^{12}$$

This can equally be

$$(a^4)^3 = a^4 \times a^4 \times a^4$$
$$= (a \times a \times a \times a) \times (a \times a \times a \times a)$$
$$\times (a \times a \times a \times a) = a^{12}$$

- **Law 4:** $a^0 = 1$

This is the law of **zero index.**

How do I arrive at this?

For example;

$$a^m \div a^m = \frac{a^m}{a^m} = 1$$
$$a^{m-m} = a^0 = 1$$

- **Law 5:** $a^{-m} = \frac{1}{a^m}$

This is a law of **Negative index.** We can still have an example to be,

$$a^{-1} = \frac{1}{a^1}$$

Just note that any time you have a **Base** having a negative index, it is equivalent to the inverse of that base and index, excluding the negative sign.

- **Law 6:** $a^{\frac{1}{m}} = \sqrt[m]{a}$

This is also known as the law of the *fractional index.*

Given an example of,

$$9^{\frac{3}{2}} = (3\times3)^{\frac{3}{2}} = (3^2)^{\frac{3}{2}} = 3^{2\times\frac{3}{2}} = 3^3 = 27$$

$$(\sqrt[2]{9})^3 = 3^3 = 27$$

- **Law 6:** $\quad a^{\frac{x}{y}} = \left(a^{\frac{1}{y}}\right)^x = (\sqrt[y]{a})^x$

For example:

$$25^{\frac{3}{2}} = \left(25^{\frac{1}{2}}\right)^3$$

$$= (\sqrt[2]{25})^3 = 5^3$$

$$= 125$$

Note the following cases:

- $(ab)^x = a^x x\ b^x$

For example

$$(2x3)^2 = 2^2x3^2 \quad = \quad 2x2x3x3$$
$$= \quad 36$$

- $\left(\frac{a}{b}\right)^x = \frac{a^x}{b^x}$

For example

$$\left(\frac{2}{3}\right)^3 = \frac{2^3}{3^3}$$

$$\frac{2x2x2}{3x3x3x} = \frac{8}{27}$$

The laws of indices given above are the basic study you have

to go through to solve any given question on indices. This is what

we will be using as we resolve miscellaneous questions on

indices in this workbook.

Chapter Two

SIMPLIFICATION OF COMPLEX INDICES

Problems or questions that involve simplification or evaluation in indices using the law of indices are reduced to the lowest form, at which they can exist.

Miscellaneous Questions and solutions

Question 1:   *Simplify* $32^{\frac{1}{5}}$

Solution
From the law of indices

$$a^{\frac{1}{m}} = \sqrt[m]{a}$$

$$32^{\frac{1}{5}} = \sqrt[5]{32}$$

$\sqrt[5]{32}$ implies a number you can multiply 5 times to give 32.

$$\sqrt[5]{32} = 2.$$

Another method

$$32^{\frac{1}{5}} = (2^5)^{\frac{1}{5}}$$

Note:

$$\mathbf{32} = 2 \times 2 \times 2 \times 2 \times 2 = 2^5$$
$$(2^5)^{\frac{1}{5}} = 2^{5 \times \frac{1}{5}}$$
$$= 2^1$$
$$= 2$$

Question 2: *Simplify* $27^{\frac{2}{3}}$

Solution

From the law of indices

$$a^{\frac{x}{y}} = \left(a^{\frac{1}{y}}\right)^x = (\sqrt[y]{a})^x$$

$$27^{\frac{2}{3}} = \left(27^{\frac{1}{3}}\right)^2 = (\sqrt[3]{27})^2$$

$\sqrt[3]{27}$ *implies a number you can multiply 3 times to give 27.*

$$\sqrt[3]{27} = 3$$

$$\left(\sqrt[3]{27}\right)^2 = 3^2 = 9$$

Another method

$$27^{\frac{2}{3}} = (3^3)^{\frac{2}{3}}$$

Note:

$$27 = 3 \times 3 \times 3 = 3^3$$

$$(3^3)^{\frac{2}{3}} = 3^{3 \times \frac{2}{3}}$$

$$= 3^2 = 9$$

Question 3: *Simplify* $\dfrac{6^{2n+1} \times 9^n \times 4^{2n}}{18^n \times 2^n \times 12^{2n}}$

Solution

$$\frac{6^{2n+1} \times 9^n \times 4^{2n}}{18^n \times 2^n \times 12^{2n}}$$

Reduce all the entities to the lowest base.

$$6^{2n+1} = 6^{2n} \times 6^1 = (2 \times 3)^{2n} \times 2 \times 3$$
$$= 2^{2n} \times 2 \times 3^{2n} \times 3$$
$$9^n = (3 \times 3)^n = (3^2)^n = 3^{\,2n}$$
$$4^{2n} = (2^2)^{2n} = 2^{4n}$$
$$18^n = (2 \times 3 \times 3)^n = (2 \times 3^2)^n = 2^n \times 3^{2n}$$
$$2^n = 2^n$$
$$12^{2n} = (2 \times 2 \times 3)^{2n} = (2^2 \times 3)^{2n} = 2^{4n} \times 3^{2n}$$

Substitute all the entity into the equations

$$= \frac{2^{2n} \times 2 \times 3^{2n} \times 3 \times 3^{\,2n} \times 2^{4n}}{2^n \times 3^{2n} \times 2^n \times 2^{4n} \times 3^{2n}}$$

$$= \frac{2^{2n} \times 2 \times 2^{4n} \times 3^{2n} \times 3 \times 3^{\,2n}}{2^n \times 2^n \times 2^{4n} \times 3^{2n} \times 3^{\,2n}}$$

$$= \frac{2^{2n+1+4n} \times 3^{2n+1+2n}}{2^{n+n+4n} \times 3^{4n}}$$

$$= \frac{2^{6n+1} \times 3^{4n+1}}{2^{6n} \times 3^{4n}}$$

$$= \frac{2^{6n} \times 2^1 \times 3^{4n} \times 3^1}{2^{6n} \times 3^{4n}}$$

$$= 2^1 \times 3^1 = 2 \times 3 = 6$$

$$= \frac{6^{2n+1} \times 9^n \times 4^{2n}}{18^n \times 2^n \times 12^{2n}} = 6$$

Question 4:   *Simplify* $4^{\frac{-3}{2}}$

Solution

From the law of indices

$$a^{\frac{x}{y}} = \left(a^{\frac{1}{y}}\right)^x = \left(\sqrt[y]{a}\right)^x$$

$$4^{\frac{-3}{2}} = \left(4^{\frac{1}{2}}\right)^{-3}$$

$$= \left(\sqrt[2]{4}\right)^{-3}$$

$\sqrt[2]{4}$ means the square root of 4

$$\sqrt[2]{4} = 2$$

$$\left(\sqrt[2]{4}\right)^{-3} = 2^{-3}$$

Remember the question given carries (-) index. –ve index means inverse.

Recall

$$a^{-1} = \frac{1}{a}$$

$$2^{-3} = \frac{1}{2^3}$$

$$\frac{1}{2^3} = \frac{1}{2\times2\times2} = \frac{1}{8}$$

$$4^{\frac{-3}{2}} = \frac{1}{8}$$

Another method

$$4^{\frac{-3}{2}} = (2^2)^{\frac{-3}{2}}$$

Note:

$$4 = 2 \times 2 = 2^2$$

$$(2^2)^{\frac{-3}{2}} = 2^{2\times\frac{-3}{2}}$$

$$= 2^{-3} = \frac{1}{2^3} = \frac{1}{8}$$

Question 5: Simplify $64^{\frac{2}{3}}$

Solution

Find the lowest common factor of 64

$$64 = 2 \times 2 \times 2 \times 2 \times 2 \times 2 = 2^6$$

$$64^{\frac{2}{3}} = (2^6)^{\frac{2}{3}}$$

$$(2^6)^{\frac{2}{3}} = 2^{6 \times \frac{2}{3}}$$

$$2^{2 \times 2} = 2^4 = 2 \times 2 \times 2 \times 2 = 16$$

$$64^{\frac{2}{3}} = 16$$

Question 6: Find the value of $(0.001)^3$

Solution

$$(0.001)^3$$

$$0.001 = \frac{1}{1000}$$

Do you know that?

$$(0.001)^3 = \left(\frac{1}{1000}\right)^3$$

Note: $10 \times 10 \times 10 = 10^3 = 1000$

$$\left(\frac{1}{1000}\right)^3 = (10^{-3})^3 = 10^{-9}$$

Note

$$a^{-x} = \frac{1}{a^x} = 10^{-9}$$

Question 7: Find the value of 8^0

Solution

$$8^0 = 1$$

Note

Any base that is raise to power of zero (o) will be equal to $1_{//}$

Question 8: **Simplify** $\sqrt{\dfrac{0.81 \times 10^{-5}}{2.25 \times 10^{7}}}$

Solution

Write in standard form

$$0.81 = 81 \times 10^{-2}$$
$$2.25 = 225 \times 10^{-2}$$

Put the standard form of the entity into the questions

$$= \sqrt{\frac{81 \times 10^{-2} \times 10^{-5}}{225 \times 10^{-2} \times 10^{7}}}$$

$$= \sqrt{\frac{81 \times 10^{-7}}{225 \times 10^{5}}}$$

$$= \sqrt{\frac{81 \times 10^{-7} \times 10^{-5}}{225}}$$

$$= \sqrt{\frac{81 \times 10^{-7} \times 10^{-5}}{225}}$$

$$= \sqrt{\frac{81 \times 10^{-12}}{225}}$$

$$= \frac{\sqrt{81} \times \sqrt{10^{-12}}}{\sqrt{225}} = \frac{9 \times \left(10^{-12 \times \frac{1}{2}} \right)}{15}$$

$$= \frac{9 \times 10^{-6}}{15}$$

$$= 0.6 \times 10^{-6}$$

$$= 6.0 \times 10^{-5}$$

Therefore;

$$\sqrt{\frac{0.81 \times 10^{-5}}{2.25 \times 10^7}} = 6.0 \times 10^{-5}$$

Question 9: *Simplify $125^{\frac{-1}{3}} \times 49^{\frac{-1}{2}} \times 8^0$*

Solution

$$125^{\frac{-1}{3}} \times 49^{\frac{-1}{2}} \times 8^0$$

Find the lowest common multiple for the bases at each side.

$$125 = 5 \times 5 \times 5 = 5^3$$

$$49 = 7 \times 7 = 7^2$$

$$8^0 = 1$$

$$(5^3)^{\frac{-1}{3}} \times (7^2)^{\frac{-1}{2}} \times 1$$

$$= 5^{3 \times \frac{-1}{3}} \times 7^{2 \times \frac{-1}{2}} \times 1$$

$$= 5^{-1} \times 7^{-1} \times 1$$

$$= \frac{1}{5} \times \frac{1}{7} = \frac{1}{35}$$

Therefore;

$$125^{\frac{-1}{3}} \times 49^{\frac{-1}{2}} \times 8^0 = \frac{1}{35}$$

Question 10: *Simplify $4^2 \times 4^{-3}$*

Solution

$$4^2 \times 4^{-3}$$

From the law of indices

$$a^m \times a^n = a^{m+n}$$

$$4^2 \times 4^{-3} = 4^{2-3}$$

$$= 4^{-1}$$

$$= \frac{1}{4}$$

$$\mathbf{4^2 \times 4^{-3}} = \frac{1}{4}$$

Question 11:   *Simplify* $\dfrac{3^2(2^2)^{-2}}{2^3}$

Solution

$$= \frac{3^2(2^2)^{-2}}{2^3}$$

$$= \frac{3^2 \times (2^2)^{-2}}{2^3}$$

$$= \frac{3^2 \times 2^{-4}}{2^3}$$

$$= 3^2 \times 2^{-4} \div 2^3$$

From the law of indices

$$a^m \div a^n = a^{m-n}$$

$$3^2 \times 2^{-4} \div 2^3$$

$$3^2 \times 2^{-4-3}$$

$$3^2 \times 2^{-7}$$

$$9 \times \frac{1}{2^7} = \frac{9}{128}$$

$$\frac{3^2(2^2)^{-2}}{2^3} = \frac{9}{128}$$

Question 12:   *Simplify* $x^5 x^8$

Solution

From the law of indices

$$a^m \times a^n = a^{m+n}$$

$$x^5 \times x^8 = x^{5+8}$$

$$= x^{13}$$

$$x^5 x^8 = x^{13}$$

Question 13: **Simplify** $(p^4)^6$

Solution

From the law of indices

$$(a^x)^y = a^{x \times y} = a^{xy}$$

$$(p^4)^6 = p^{4 \times 6}$$

$$= p^{24}$$

$$(p^4)^6 = p^{24}$$

Question 14: **Simplify** (-3^3)

Solution

Note

$$+ \times + = +$$

$$- \times - = +$$

$$+ \times - = -$$

$$- \times + = -$$

$$(-3) \times (-3) = 9$$

$$9 \times (-3) = -27$$

$$(-3^3) = -27$$

$$(-3^3) = -27$$

Question 15: *Simplify* $(4ab^2c)^3$

Solution

$$(4ab^2c)^3$$

From the law of indices

$$a^m \times a^n = a^{m+n}$$

$$= 4^3 \times a^3 \times (b^2)^3 \times c^3$$

$$= 4 \times 4 \times 4 \times a^3 \times b^{2 \times 3} \times c^3$$

$$= 64a^3b^6c^3$$

$$(4ab^2c)^3 = 64a^3b^6c^3$$

Question 16: *Simplify* $x^2z^{-3} \times (xz^2)^2$

Solution

$$= x^2z^{-3} \times (xz^2)^2$$

$$= x^2 \times \frac{1}{z^3} \times x^2 \times z^{2 \times 2}$$

From the law of indices

$$a^m \times a^n = a^{m+n}$$

Collect like terms together

$$= x^2 \times x^2 \times z^{-3} \times z^4$$

$$= x^{2+2} \times z^{-3+4}$$

$$= x^4 \times z^1 = x^4z$$

$$x^2z^{-3} \times (xz^2)^2 = x^4z$$

Question 17:   *Simplify $2^n \times (2^{-n})^3 \times 2^{2n}$*

Solution

$$2^n \times (2^{-n})^3 \times 2^{2n}$$

$$= 2^n \times 2^{-n \times 3} \times 2^{2n}$$

$$= 2^n \times 2^{-3n} \times 2^{2n}$$

$$= 2^{n+(-3n)+2n}$$

$$= 2^{n-3n+2n}$$

$$= 2^0 = 1$$

$$2^n \times (2^{-n})^3 \times 2^{2n} = 1$$

Question 18:   *Simplify $3^m \times 27^m \times 9^{-m}$*

Solution

$$3^m \times 27^m \times 9^{-m}$$

Find the lowest common multiples for all the bases.

$$3 = 3^1$$

$$9 = 3 \times 3 = 3^2$$

$$27 = 3 \times 3 \times 3 = 3^3$$

$$3^m \times 27^m \times 9^{-m}$$

$$3^m \times (3^3)^m \times (3^2)^{-m}$$

$$3^m \times 3^{3m} \times 3^{-2m}$$

$$3^{m+3m-2m}$$

$$3^{4m-2m}$$

$$3^{2m} = 9^m$$

$$3^m \times 27^m \times 9^{-m} = 9^m$$

Question 19:   *Simplify* $\left(a^{\frac{1}{2}} \times a\right)^5$

<u>Solution</u>

$$\left(a^{\frac{1}{2}} \times a\right)^5$$

$$= (a^{\frac{1}{2}+1})^5$$

$$= \left(a^{\frac{3}{2}}\right)^5$$

$$= a^{\frac{15}{2}} = \left(\sqrt{a}\right)^{15}$$

$$\left(a^{\frac{1}{2}} \times a\right)^5 = a^{\frac{15}{2}} = \left(\sqrt{a}\right)^{15}$$

Question 20:   *Simplify* $\left(\frac{-2ab}{2b}\right)^2$

<u>Solution</u>

$$\left(\frac{-2ab}{2b}\right)^2 = \frac{(-2)^2 \times a^2 \times b^2}{2 \times b}$$

$$= \frac{4 \times a^2 \times b^2 \times b^{-1}}{2}$$

$$= \frac{4 \times a^2 \times b^{2-1}}{2}$$

$$= \frac{4 \times a^2 \times b^1}{2}$$

$$= \frac{4a^2b}{2} = 2a^2b$$

$$= 2a^2b$$

$$\left(\frac{-2ab}{2b}\right)^2 = 2a^2b$$

Question 21:   *Simplify* $\dfrac{\left(-a^4b\right)^3 (ab)^5}{-a^8b^8}$

Solution

$$\frac{-a^{4\times3}b^3 a^5 b^5}{-a^8 b^8}$$

$$= \frac{-a^{12}b^3 a^5 b^5}{-a^8 b^8}$$

$$= \frac{a^{12+5} \times b^{3+5}}{a^8 b^8}$$

$$= \frac{a^{17} b^8}{a^8 b^8}$$

$$= a^{17-8} b^{8-8}$$

$$= a^9 b^0$$

$$= a^9 \times 1$$

$$= a^9$$

$$\frac{\left(-a^4b\right)^3 (ab)^5}{-a^8 b^8} = a^9$$

Question 22:   *Simplify* $\dfrac{x^{-1}y^4}{x^{-5}y^{-3}}$

Solution

$$\frac{x^{-1}y^4}{x^{-5}y^{-3}}$$

From the laws of indices

$$a^m \div a^n = a^{m-n}$$

$$\frac{x^{-1}y^4}{x^{-5}y^{-3}}$$

$$= (x^{-1} \div x^{-5}) \times (y^4 \div y^{-3})$$

$$= x^{-1-(-5)} \times y^{4-(-3)}$$

$$= x^{-1+5} \times y^{4+3}$$

$$= x^4 \times y^7$$

$$= x^4 y^7$$

$$\frac{x^{-1}y^4}{x^{-5}y^{-3}} = x^4 y^7$$

Question 23:   **Simplify** $\left(\frac{10a^3 b^{-2}}{5a^{-1}b^2}\right)^{-1}$

<u>**Solution**</u>

$$\left(\frac{10a^3 b^{-2}}{5a^{-1}b^2}\right)^{-1}$$

$$= \left(\frac{10a^3 b^{-2}}{5a^{-1}b^2}\right)^{-1}$$

$$= \left(\frac{10}{5} \times a^3 \div a^{-1} \times b^{-2} \div b^2\right)^{-1}$$

From the law of indices

$$= a^m \div a^n = a^{m-n}$$

$$= (2 \times a^3 \div a^{-1} \times b^{-2} \div b^2)^{-1}$$

$$= (2 \times a^3 \div a^{-1} \times b^{-2} \div b^2)^{-1}$$

$$= \left(2 \times a^{3-(-1)} \times b^{-2-2}\right)^{-1}$$

$$= (2 \times a^{3+1} \times b^{-4})^{-1}$$

$$= (2 \times a^4 \times b^{-4})^{-1}$$

$$= (2a^4 b^{-4})^{-1}$$

Negative index gives an inverse

$$(2a^4 b^{-4})^{-1} = \frac{1}{2a^4 b^{-4}}$$

$$\left(\frac{10a^3 b^{-2}}{5a^{-1} b^2}\right)^{-1} = \frac{1}{2a^4 b^{-4}}$$

Question 24: Simplify $x\sqrt[3]{x}$.

Solution

$$x\sqrt[3]{x} = x^1 \times x^{\frac{1}{3}}$$

$$= x^{1+\frac{1}{3}}$$

$$= x^{\frac{2}{3}}$$

$$= \left(\sqrt[3]{x}\right)^2$$

$$x\sqrt[3]{x} = \left(\sqrt[3]{x}\right)^2$$

Question 25: Simplify $\left(a^2 \times \sqrt{a}\right)^2$

Solution

$$\left(a^2 \times \sqrt{a}\right)^2$$

$$\left(a^2 \times a^{\frac{1}{2}}\right)^2$$

$$\left(a^{2+\frac{1}{2}}\right)^2$$

$$\left(a^{\frac{5}{2}}\right)^2$$

$$a^{\frac{5}{2} \times 2}$$

$$= a^5$$

$$\left(a^2 \times \sqrt{a}\right)^2 = a^5$$

Question 26: *Simplify $\frac{2x^{\frac{1}{2}}x}{x^2}$*

Solution

$$\frac{2x^{\frac{1}{2}}x}{x^2} = \frac{2 \times x^{\frac{1}{2}} \times x}{x^2}$$

$$= \frac{2x^{\frac{1}{2}+1}}{x^2} = \frac{2x^{\frac{3}{2}}}{x^2}$$

$$= 2x^{\frac{3}{2}} \div x^2 = 2x^{\frac{3}{2}-2}$$

$$2x^{\frac{-1}{2}} = \frac{2}{\sqrt{x}}$$

$$\frac{2x^{\frac{1}{2}}x}{x^2} = \frac{2}{\sqrt{x}}$$

Question 27: *Simplify $(3a)^{-1} \times 3a^{-1}$*

Solution

$$(3a)^{-1} \times 3a^{-1}$$

Negative index means an inverse.

$$\frac{1}{3a} \times \frac{3}{a} = \frac{3}{3a^2}$$

$$= \frac{1}{a^2}$$

$$(3a)^{-1} \times 3a^{-1} = \frac{1}{a^2}$$

Question 28:   *Simplify $32^{\frac{3}{5}}$*

Solution

$$32^{\frac{3}{5}}$$

Find the lowest common multiple of the base

$$32 = 2 \times 2 \times 2 \times 2 \times 2 = 2^5$$

$$32^{\frac{3}{5}} = (2^5)^{\frac{3}{5}}$$

$$= 2^{5 \times \frac{3}{5}} = 2^3$$

$$= 2^3 = 8$$

$$32^{\frac{3}{5}} = 8$$

Question 29:   *Simplify $\left(\frac{4}{25}\right)^{\frac{3}{2}}$*

Solution
Find the lowest common multiple for the bases.

$$4 = 2 \times 2 = 2^2$$

$$25 = 5 \times 5 = 5^2$$

$$\left(\frac{4}{25}\right)^{\frac{3}{2}} = \left(\frac{2 \times 2}{5 \times 5}\right)^{\frac{3}{2}}$$

$$\left(\frac{2^2}{5^2}\right)^{\frac{3}{2}} = \left(\frac{2}{5}\right)^{2\times\frac{3}{2}}$$

$$\left(\frac{2}{5}\right)^3 = \frac{2\times2\times2}{5\times5\times5} = \frac{8}{125}$$

$$\left(\frac{4}{25}\right)^{\frac{3}{2}} = \frac{8}{125}$$

Question 30:   *Simplify* $\left(4^{\frac{1}{3}}\right)\left(2^{\frac{1}{3}}\right)$

Solution
Find the lowest common multiple for the bases.

$$2 = 2^1$$

$$4 = 2 \times 2 = 2^2$$

$$\left(4^{\frac{1}{3}}\right)\left(2^{\frac{1}{3}}\right) = (2^2)^{\frac{1}{3}} \times 2^{\frac{1}{3}}$$

$$\left(2^{2\times\frac{1}{3}}\right)\left(2^{\frac{1}{3}}\right) = \left(2^{\frac{2}{3}}\right)\left(2^{\frac{1}{3}}\right)$$

$$2^{\frac{2}{3}+\frac{1}{3}} = 2^{\frac{3}{3}} = 2^1$$

$$= 2$$

$$\left(4^{\frac{1}{3}}\right)\left(2^{\frac{1}{3}}\right) = 2$$

Question 31:   *Evaluate* $\frac{81^{\frac{3}{4}}-27^{\frac{1}{3}}}{3\times2^3}$

Solution

$$\frac{81^{\frac{3}{4}}-27^{\frac{1}{3}}}{3\times2^3}$$

$$= \frac{(3^4)^{\frac{3}{4}}-(3^3)^{\frac{1}{3}}}{3\times2^3}$$

$$= \frac{3^{4 \times \frac{3}{4}} - 3^{3 \times \frac{1}{3}}}{3 \times 2^3}$$

$$= \frac{3^3 - 3^1}{3 \times 8}$$

$$= \frac{27 - 3}{24}$$

$$= \frac{24}{24} = 1$$

$$\frac{81^{\frac{3}{4}} - 27^{\frac{1}{3}}}{3 \times 2^3} = 1$$

Question 32:

Simplify $(25)^{\frac{-1}{2}} \times (27)^{\frac{1}{3}} + (121)^{\frac{-1}{2}} \times (625)^{\frac{-1}{4}}$

Solution

$$(25)^{\frac{-1}{2}} \times (27)^{\frac{1}{3}} + (121)^{\frac{-1}{2}} \times (625)^{\frac{-1}{4}}$$

The lowest common multiple for all the bases are;

$$25 = 5 \times 5 = \mathbf{5^2}$$

$$27 = 3 \times 3 \times 3 = \mathbf{3^3}$$

$$121 = 11 \times 11 = \mathbf{11^2}$$

$$625 = 5 \times 5 \times 5 \times 5 = \mathbf{5^4}$$

$$= (25)^{\frac{-1}{2}} \times (27)^{\frac{1}{3}} + (121)^{\frac{-1}{2}} \times (625)^{\frac{-1}{4}}$$

$$= (5^2)^{\frac{-1}{2}} \times (3^3)^{\frac{1}{3}} + (11^2)^{\frac{-1}{2}} \times (5^4)^{\frac{-1}{4}}$$

$$= 5^{-1} \times 3^1 + 11^{-1} \times 5^{-1}$$

$$= \frac{1}{5} \times 3 + \frac{1}{11} \times \frac{1}{5}$$

$$= \frac{3}{5} + \frac{1}{55} = \frac{33 + 1}{55} = \frac{34}{55}$$

$$(25)^{\frac{-1}{2}} \times (27)^{\frac{1}{3}} + (121)^{\frac{-1}{2}} \times (625)^{\frac{-1}{4}} = \frac{34}{55}$$

Question 33: *Simplify* $16^{\frac{-1}{2}} \times 4^{\frac{-1}{2}} \times 27^{\frac{1}{3}}$

Solution

$$16^{\frac{-1}{2}} \times 4^{\frac{-1}{2}} \times 27^{\frac{1}{3}}$$

Find the lowest common multiple of the bases

$$16 = 2 \times 2 \times 2 \times 2 = 2^4$$
$$4 = 2 \times 2 = 2^2$$
$$27 = 3 \times 3 \times 3 = 3^3$$

Substitute the lowest common multiple forms of those bases into the equation

$$= (2^4)^{\frac{-1}{2}} \times (2^2)^{\frac{-1}{2}} \times (3^3)^{\frac{1}{3}}$$
$$= 2^{4 \times \frac{-1}{2}} \times 2^{2 \times \frac{-1}{2}} \times 3^{3 \times \frac{1}{3}}$$
$$= 2^{-2} \times 2^{-1} \times 3^1$$
$$= \frac{1}{2^2} \times \frac{1}{2^1} \times 3$$
$$= \frac{1}{4} \times \frac{1}{2} \times 3$$
$$= \frac{3}{8}$$

$$16^{\frac{-1}{2}} \times 4^{\frac{-1}{2}} \times 27^{\frac{1}{3}} = \frac{3}{8}$$

Question 34: *Simplify* $\left(\frac{81}{16}\right)^{\frac{-1}{4}} \times 2^{-1}$

Solution

$$\left(\frac{81}{16}\right)^{\frac{-1}{4}} \times 2^{-1}$$

$$\left(\frac{3^4}{2^4}\right)^{\frac{-1}{4}} \times 2^{-1}$$

$$\left(\left(\frac{3}{2}\right)^4\right)^{\frac{-1}{4}} \times \frac{1}{2} = \left(\frac{3}{2}\right)^{-1} \times \frac{1}{2}$$

$$\frac{2}{3} \times \frac{1}{2} = \frac{1}{3}$$

$$\left(\frac{81}{16}\right)^{\frac{-1}{4}} \times 2^{-1} = \frac{1}{3}$$

Question 35:   *Simplify $\left(\frac{16}{81}\right)^{\frac{1}{4}} \div \left(\frac{9}{16}\right)^{\frac{-1}{2}}$*

Solution

$$\left(\frac{16}{81}\right)^{\frac{1}{4}} \div \left(\frac{9}{16}\right)^{\frac{-1}{2}}$$

Find the lowest common multiple for all the bases.

$$16 = 2 \times 2 \times 2 \times 2 = 2^4$$
$$81 = 3 \times 3 \times 3 \times 3 = 3^4$$
$$9 = 3 \times 3 \times 3 = 3^3$$

Substitute into the equations

$$\left(\frac{2^4}{3^4}\right)^{\frac{1}{4}} \div \left(\frac{3^2}{4^2}\right)^{\frac{-1}{2}}$$

$$\left(\left(\frac{2}{3}\right)^4\right)^{\frac{1}{4}} \div \left(\left(\frac{3}{4}\right)^2\right)^{\frac{-1}{2}}$$

$$\left(\frac{2}{3}\right)^{4\times\frac{1}{4}} \div \left(\frac{3}{4}\right)^{2\times\frac{-1}{2}}$$

$$\left(\frac{2}{3}\right)^1 \div \left(\frac{3}{4}\right)^{-1}$$

$$\frac{2}{3} \div \frac{1}{\frac{3}{4}} = \frac{2}{3} \div \frac{4}{3}$$

$$= \frac{2}{3} \times \frac{3}{4} = \frac{6}{12} = \frac{1}{2}$$

$$\left(\frac{16}{81}\right)^{\frac{1}{4}} \div \left(\frac{9}{16}\right)^{\frac{-1}{2}} = \frac{1}{2}$$

Question 36:   Simplify $(0.00001)^2$

Solution

0.00001 into fraction $= \dfrac{1}{100000} = \dfrac{1}{10^5} = 10^{-5}$

$(0.00001)^2 = (10^{-5})^2$

$= 10^{-5\times2} = 10^{-10}$

$$(0.00001)^2 = 10^{-10}$$

Question 37:   Simplify $343^{\frac{2}{3}}$

Solution

Find the lowest common multiple of the base 343

$$343 = 7 \times 7 \times 7 = 7^3$$

$$343^{\frac{2}{3}} = (7^3)^{\frac{2}{3}}$$

$$7^{3 \times \frac{2}{3}} = 7^2 = 7 \times 7 = 49$$

$$343^{\frac{2}{3}} = 49$$

Question 38:   Simplify $216^{\frac{4}{3}}$

Solution

Find the lowest common multiple of the base 216

$$216 = 6 \times 6 \times 6 = 6^3$$

$$216^{\frac{4}{3}} = (6^3)^{\frac{4}{3}}$$

$$= 6^{3 \times \frac{4}{3}} = 6^4$$

$$= 6 \times 6 \times 6 \times 6 = 1296$$

$$216^{\frac{4}{3}} = 1296$$

Question 39:   Simplify $\dfrac{5^a \times 5^{a-1}}{125^{a+1}}$

Solution

$$\frac{5^a \times 5^{a-1}}{125^{a+1}}$$

Find the lowest common base for bases.

$$\frac{5^a \times 5^{a+1}}{125^{a+1}}$$

$$125 = 5 \times 5 \times 5 = 5^3$$

$$\frac{5^a \times 5^{a-1}}{(5^3)^{a+1}} = \frac{5^a \times 5^{a-1}}{5^{3(a+1)}}$$

$$= \frac{5^a \times 5^{a-1}}{5^{3a+3}}$$

From the laws of indices.

$$5^a \times 5^{a-1} = 5^{a+a-1}$$

$$= 5^{2a-1}$$

$$\frac{5^{2a-1}}{5^{3a+3}}$$

From $\quad \dfrac{a^m}{a^n} = a^{m-n}$

$$5^{2a-1} \div 5^{3a-3}$$

$$5^{2a-1-(3a+3)}$$

$$5^{2a-1-3a-3}$$

$$5^{2a-3a-1-3}$$

$$5^{-a-4}$$

$$5^{-(a+4)} = \frac{1}{5^{a+4}}$$

$$\frac{5^a \times 5^{a-1}}{125^{a+1}} = \frac{1}{5^{a+4}}$$

Question 40: Simplify $25^{\frac{1}{2}} \times 8^{\frac{-2}{3}}$

Solution

$$25^{\frac{1}{2}} \times 8^{\frac{-2}{3}}$$

Find the lowest common multiple for the bases at each side.

$$25 = 5 \times 5 = 5^2$$

$$8 = 2 \times 2 \times 2 = 2^3$$

$$25^{\frac{1}{2}} \times 8^{\frac{-2}{3}}$$

$$(5^2)^{\frac{1}{2}} \times (2^3)^{\frac{-2}{3}}$$

$$5^{2\times\frac{1}{2}} \times 2^{3\times\frac{-2}{3}}$$

$$5^1 \times 2^{-2}$$

$$5 \times \frac{1}{2^2}$$

$$5 \times \frac{1}{4} = \frac{5}{4} = 1\frac{1}{4}$$

$$25^{\frac{1}{2}} \times 8^{\frac{-2}{3}} = 1\frac{1}{4}$$

Question 41: *Simplify* $25^{1.5}$

Solution

$$1.5 \text{ to fraction} = \frac{15}{10} = \frac{3}{2}$$

$$25^{1.5} = 25^{\frac{3}{2}}$$

Find the lowest common multiple for the base

$$25 = 5 \times 5 = 5^2$$

$$25^{\frac{3}{2}} = (5^2)^{\frac{3}{2}}$$

$$= 5^{2\times\frac{3}{2}} = 5^3$$

$$= 125$$

$$25^{1.5} = 125$$

Question 42: *Simplify* $\frac{3^n - 3^{n-1}}{3^3 \times 3^n - 27 \times 3^{n-1}}$

Solution

$$\frac{3^n - 3^{n-1}}{3^3 \times 3^n - 3^3 \times 3^{n-1}} = \frac{3^n - 3^{n-1}}{3^3(3^n - 3^{n-1})}$$

$$= \frac{1}{3^3} = \frac{1}{27}$$

$$\frac{3^n - 3^{n-1}}{3^3 \times 3^n - 27 \times 3^{n-1}} = \frac{1}{27}$$

Question 43: *Simplify* $\dfrac{9^{\frac{1}{3}} \times 27^{\frac{-1}{2}}}{3^{\frac{-1}{6}} \times 3^{\frac{-2}{3}}}$

Solution

$$\frac{9^{\frac{1}{3}} \times 27^{\frac{-1}{2}}}{3^{\frac{-1}{6}} \times 3^{\frac{-2}{3}}}$$

Reduce the bases to their lowest common base

$$9 = 3 \times 3 = 3^2$$

$$27 = 3 \times 3 \times 3 = 3^3$$

$$= \frac{(3^2)^{\frac{1}{3}} \times (3^3)^{\frac{-1}{2}}}{3^{\frac{-1}{6}} \times 3^{\frac{-2}{3}}}$$

$$= \frac{3^{2 \times \frac{1}{3}} \times 3^{3 \times \frac{-1}{2}}}{3^{\frac{-1}{6}} \times 3^{\frac{-2}{3}}} = \frac{3^{\frac{2}{3}} \times 3^{\frac{-3}{2}}}{3^{\frac{-1}{6}} \times 3^{\frac{-2}{3}}}$$

$$= \frac{3^{\frac{2}{3} + \left(-\frac{3}{2}\right)}}{3^{\frac{-1}{6} + \left(-\frac{2}{3}\right)}}$$

$$= \frac{3^{\frac{2}{3} - \frac{3}{2}}}{3^{-\frac{1}{6} - \frac{2}{3}}}$$

$$= \frac{3^{\frac{4-9}{6}}}{3^{\frac{-1-4}{6}}}$$

$$= \frac{3^{\frac{-5}{6}}}{3^{\frac{-5}{6}}} = 1$$

$$\frac{9^{\frac{1}{3}} \times 27^{\frac{-1}{2}}}{3^{\frac{-1}{6}} \times 3^{\frac{-2}{3}}} = 1$$

Question 44:

Evaluate without using the tables $(0.008)^{\frac{-1}{3}} \times (0.16)^{\frac{-3}{2}}$

Solution

$$(0.008)^{\frac{-1}{3}} \times (0.16)^{\frac{-3}{2}}$$

Change the bases to a fraction

$$0.008 = \frac{8}{1000}$$

$$0.16 = \frac{16}{100}$$

$$\left(\frac{8}{1000}\right)^{\frac{-1}{3}} \times \left(\frac{16}{100}\right)^{\frac{-3}{2}}$$

$$\left(\frac{2 \times 2 \times 2}{10 \times 10 \times 10}\right)^{\frac{-1}{3}} \times \left(\frac{2 \times 2 \times 2}{10 \times 10 \times 10}\right)^{\frac{-3}{2}}$$

$$\left(\frac{2^3}{10^3}\right)^{\frac{-1}{3}} \times \left(\frac{4^2}{10^2}\right)^{\frac{-3}{2}}$$

$$\left(\frac{2}{10}\right)^{3 \times \frac{-1}{3}} \times \left(\frac{4}{10}\right)^{2 \times \frac{-3}{2}}$$

$$= \left(\frac{2}{10}\right)^{-1} \times \left(\frac{4}{10}\right)^{-3}$$

$$= \left(\frac{1}{5}\right)^{-1} \times \left(\frac{2}{5}\right)^{-3}$$

$$= \frac{1}{\left(\frac{1}{5}\right)^1} \times \frac{1}{\left(\frac{2}{5}\right)^3}$$

$$= \frac{1}{\frac{1}{5}} \times \frac{1}{\frac{8}{125}}$$

$$= \frac{5}{1} \times \frac{125}{8}$$

$$= \frac{625}{8}$$

$$(0.008)^{\frac{-1}{3}} \times (0.16)^{\frac{-3}{2}} = \frac{625}{8}$$

Question 45: *Evaluate* $\dfrac{8^{\frac{1}{3}} \times 5^{\frac{2}{3}}}{10^{\frac{2}{3}}}$

Solution

$$\frac{8^{\frac{1}{3}} \times 5^{\frac{2}{3}}}{10^{\frac{2}{3}}} = \frac{(2 \times 2 \times 2)^{\frac{1}{3}} \times 5^{\frac{2}{3}}}{(2 \times 5)^{\frac{2}{3}}}$$

$$= \frac{(2^3)^{\frac{1}{3}} \times 5^{\frac{2}{3}}}{2^{\frac{2}{3}} \times 5^{\frac{2}{3}}}$$

$$= \frac{2^1}{2^{\frac{2}{3}}} = 2^{1 - \frac{2}{3}}$$

$$= 2^{\frac{3-2}{3}} = 2^{\frac{1}{3}} = \sqrt[3]{2}$$

$$\frac{8^{\frac{1}{3}} \times 5^{\frac{2}{3}}}{10^{\frac{2}{3}}} = \sqrt[3]{2}$$

Question 46: Without using tables, evaluate $343^{\frac{1}{3}} \times 0.14^{-1}$

Solution
The lowest common multiple the base

$$343 = 7 \times 7 \times 7 = 7^3$$

$$0.14 = \frac{14}{100} = \frac{7}{50}$$

$$343^{\frac{1}{3}} \times 0.14^{-1}$$

$$= (7^3)^{\frac{1}{3}} \times \frac{1}{0.14}$$

$$= 7 \times \frac{1}{\frac{7}{50}}$$

$$= 7 \times \frac{50}{7}$$

$$= 50$$

$$343^{\frac{1}{3}} \times 0.14^{-1} = 50$$

EXERCISE 1

Simplify the following.

1. $343^{\frac{4}{3}}$

2. $7x \times 5x^3$

3. $\dfrac{32x^5}{8x^2}$

4. 5^0

5. $\left(\frac{3}{4}\right)^{-1}$

6. $7x^2 \times 3x^0 \times 4x^{-2}$

7. $\dfrac{65a^2b^{-2}}{5a^3b^{-3}}$

8. $(0.125)^{\frac{1}{3}}$

9. $\sqrt[4]{16a^{-12}}$

10. $\left(\dfrac{18}{32}\right)^{\frac{-3}{2}}$

Chapter Three

ALGEBRAIC INDICES PROBLEM

These are indices problems in algebra form. They involve solving for a value that is represented with a letter.

Miscellaneous Questions and solutions

Question 47: **Solve the equation** $(5^2)^{x-1} \times 5^{x+1} = 0.04$

Solution

$$(5^2)^{x-1} \times 5^{x+1} = 0.04$$

$$0.04 \text{ into fraction} = \frac{4}{100}$$

$$5^{2x-2} \times 5^{x+1} = \frac{4}{100}$$

From the law of indices

$$a^m \times a^n = a^{m+n}$$

$$5^{2x-2+x+1} = \frac{1}{25}$$

$$5^{2x+x+1-2} = \frac{1}{5^2}$$

$$5^{3x-1} = 5^{-2}$$

Since the bases on both sides are the same, they cancel out.

$$3x - 1 = -2$$

$$3x = -2 + 1$$

$$3x = -1$$

Divide both sides by 3

$$\frac{3x}{3} = \frac{-1}{3}$$

$$x = \frac{-1}{3}$$

Question 48: Given that $\sqrt[3]{4^{2x}} = 16$. Find the value of x.

Solution

$$\sqrt[3]{4^{2x}} = 16$$

$$4^{\frac{2x}{3}} = 4 \times 4$$

$$4^{\frac{2x}{3}} = 4^2$$

Since the bases at both sides are the same, it will cancel out.

$$\frac{2x}{3} = 2$$

Cross multiply

$$2x = 2 \times 3$$

$$2x = 6$$

Divide both sides by 2

$$\frac{2x}{2} = \frac{6}{2}$$

$$x = 3$$

Question 49: Find the value of x; $(½)^x = 8$

Solution

$$(½)^x = 8$$

From the law of indices

$$\frac{1}{a} = a^{-1}$$

$$(½)^x = 8$$

$$(2^{-1})^x = 2 \times 2 \times 2$$

$$(2^{-1})^x = 2^3$$

$$2^{-1 \times x} = 2^3$$

$$2^{-x} = 2^3$$

$$-x = 3$$

Multiply through by " - "

$$x = -3$$

Question 50:   **Solve for x; 3(3)^x = 27**

Solution
From the law of indices

$$3^1 = 3$$

$$3^1 \times 3^x = 27$$

$$3^{1+x} = 3 \times 3 \times 3 = 3^3$$

$$3^{1+x} = 3^3$$

$$1+x = 3$$

$$x = 3-1$$

$$x = 2_{//}$$

Question 51:   **Solve for x; (0.25)$^{x+1}$ = 16**

Solution
Convert 0.25 to a fraction

$$\frac{25}{100} = \frac{1}{4}$$

$$(\frac{1}{4})^{x+1} = 16$$

$$(½ × 2)^{x+1} = 2×2×2×2$$

$$(½^2)^{x+1} = 2^4$$

$$(2^{-2})^{x+1} = 2^4$$

$$2^{-2(x+1)} = 2^4$$

$$2^{-2x+2} = 2^4$$

$$-2x-2 = 4$$

$$-2x = 4+2$$

$$-2x = 6$$

$$x = 6/-2$$

$$x = -3 //$$

Question 52:   Solve for x; $3^x = \dfrac{1}{81}$

Solution

Note:

$$81 = 3×3×3×3$$
$$= 3^4$$

From law of indices

$$a^{-m} = \frac{1}{a^m}$$

$$3^x = \frac{1}{3^4}$$
$$3^x = 3^{-4}$$

Since the bases at both sides are the same, it will counsel out.

$$x = -4$$

Question 53:   Solve for x; $10^{-x} = 0.0001$

Solution

$$0.0001 = \frac{1}{10000}$$

N.B

$$10000 = 10 \times 10 \times 10 \times 10 = 10^4$$

$$10^{-x} = 0.0001$$
$$10^{-x} = \frac{1}{10^4}$$

From the law of indices

$$a^{-m} = \frac{1}{a^m}$$

$$10^{-x} = 10^{-4}$$

Since the bases at both sides are the same, it will cancel out.

$$-x = -4$$
$$x = 4$$

Question 54:   **Solve for x in** $9^{x-1} = 27^{x+1}$

Solution

Find the lowest common multiple for the bases at each side

$$9 = 3 \times 3 = 3^2$$

$$27 = 3 \times 3 \times 3 = 3^3$$

$$9^{x-1} = 27^{x+1}$$

$$(3^2)^{x-1} = (3^3)^{x+1}$$

$$3^{2(x-1)} = 3^{3(x+1)}$$

Since the bases at both sides are the same, it will cancel out.

$$2(x-1) = 3(x+1)$$

$$2(x) - 2(1) = 3(x) + 3(1)$$

$$2x - 2 = 3x + 3$$

Combine like terms

$$2x - 3x = 3 + 2$$

$-x = 5$

$$x = -5$$

Question 55: If $8^{\frac{x}{2}} = 2^{\frac{3}{8}} \times 4^{\frac{3}{4}}$. Find x.

Solution
Find the lowest common multiple for the bases at each side

$$8 = 2 \times 2 \times 2 = 2^3$$

$$4 = 2 \times 2 = 2^2$$

$$2 = 2 = 2^1$$

$$8^{\frac{x}{2}} = 2^{\frac{3}{8}} \times 4^{\frac{3}{4}}$$

$$(2^3)^{\frac{x}{2}} = 2^{\frac{3}{8}} \times (2^2)^{\frac{3}{4}}$$

$$2^{3 \times \frac{x}{2}} = 2^{\frac{3}{8}} \times 2^{2 \times \frac{3}{4}}$$

$$2^{\frac{3x}{2}} = 2^{\frac{3}{8}} \times 2^{\frac{6}{4}}$$

From the law of indices

$$a^m \times a^n = a^{m+n}$$

$$2^{\frac{3x}{2}} = 2^{\frac{3}{8}+\frac{6}{4}}$$

$$2^{\frac{3x}{2}} = 2^{\frac{30}{16}}$$

Since the bases at both sides are the same, it will cancel out.

$$\frac{3x}{2} = \frac{30}{16}$$

Cross multiply

$$3x \times 16 = 2 \times 30$$

$$48x = 60$$

Divide through by 48

$$\frac{48x}{48} = \frac{60}{48}$$

$$x = \frac{5}{4}$$

Question 56:   *Solve for x in* $9^{2x+1} = \dfrac{81^{x-2}}{3^x}$

Solution
Find the lowest common multiple for the bases at each side

$81 = 3 \times 3 \times 3 \times 3 = 3^4$

$9 = 3 \times 3 = 3^2$

$3 = 3^1$

$(3^2)^{2x+1} = \dfrac{(3^4)^{x-2}}{3^x}$

$3^{(2)2x+1} = \dfrac{3^{(4)x-2}}{3^x}$

$3^{4x+2} = \dfrac{3^{4x-8}}{3^x}$

Cross multiply

$3^{4x+2} \times 3^x = 3^{4x-8}$

From the law of indices

$a^m \times a^n = a^{m+n}$

$3^{4x+2+x} = 3^{4x-8}$

$3^{5x+2} = 3^{4x-8}$

Since the bases at both sides are the same, it will cancel out

$5x + 2 = 4x - 8$

Bring liked terms together

$5x - 4x = -8 - 2$

$$x = -10$$

Question 57: **Solve for x in** $2^{2x+1} = 64$

Solution
Find the lowest common multiple for the bases at each side.

$$64 = 2 \times 2 \times 2 \times 2 \times 2 \times 2 = 2^6$$

$$2^{2x+1} = 2^6$$

Since the bases are the same, then they will cancel.

$$2x + 1 = 6$$

$$2x = 6 - 1$$

$$2x = 5$$

Divide both sides by 2

$$\frac{2x}{2} = \frac{5}{2}$$

$$x = \frac{5}{2}$$

Question 58: Given that; $\frac{1}{8^{2-3y}} = 2^{y+2}$, **find the value of y.**

Solution
From the law of indices

$$a^{-m} = \frac{1}{a^m}$$

$$8^{-(2-3y)} = 2^{y+2}$$

Find a common base for both sides

$$8 = 2 \times 2 \times 2 = 2^3$$

$$2 = 2^1$$

$$(2^3)^{-(2-3y)} = 2^{y+2}$$

$$2^{-(6-9y)} = 2^{y+2}$$

$$2^{-6+9y} = 2^{y+2}$$

Since the bases are the same, then they will cancel.

$$-6 + 9y = y + 2$$

Bring like terms together

$$9y - y = 2 + 6$$

$$8y = 8$$

Divide through by 8

$$\frac{8y}{8} = \frac{8}{8}$$

$$y = 1.$$

Question 59:   Given that; $2^a = 0.125$, find the value of a.

<u>Solution</u>

$$2^a = 0.125$$

$$0.125 \text{ to fraction} = \frac{125}{1000} = \frac{1}{8} = 8^{-1}$$

Lowest common multiple for 8

$$8 = 2 \times 2 \times 2 = 2^3$$

$$2^a = 8^{-1}$$

$$2^a = (2^3)^{-1}$$

$$2^a = 2^{-3}$$

Since the bases at both sides are the same, they will cancel.

$$a = -3.$$

Question 60:   **Solve the equation** $3^x = 81$

Solution
Find the lowest common multiple for the base

$$81 = 3 \times 3 \times 3 \times 3 = 3^4$$

$$3^x = 81$$

$$3^x = 3^4$$

Since the bases on both sides are the same, then both bases will cancel.

$$x = 4.$$

Question 61:   **Solve the equation** $8^x = 0.25$

Solution

$$8^x = 0.25$$

$$0.25 \text{ in fraction} = \frac{25}{100} = \frac{1}{4} = 4^{-1}$$

$$8^x = 4^{-1}$$

Find the lowest common multiple for the bases on the sides.

$$8 = 2 \times 2 \times 2 = 2^3$$

$$4 = 2 \times 2 = 2^2$$

$$8^x = 4^{-1}$$

$$(2^3)^x = (2^2)^{-1}$$

$$2^{3 \times x} = 2^{-2}$$

$$2^{3x} = 2^{-2}$$

Since both sides have the same base, then both bases will cancel.

$$3x = -2$$

Divide both sides by 3

$$\frac{3x}{3} = \frac{-2}{3}$$

$$x = \frac{-2}{3}$$

Question 62:   *Solve the equation;* $9^a = \frac{1}{729}$

<u>Solution</u>

$$9^a = \frac{1}{729}$$

Find the lowest common multiple for the bases

$$9 = 3 \times 3 = 3^2$$

$$729 = 3 \times 3 \times 3 \times 3 \times 3 = 3^6$$

Therefore;

$$9^a = \frac{1}{729}$$

$$(3^2)^a = \frac{1}{3^6}$$

Note: $\frac{1}{3^6} = 3^{-6}$

$$(3^2)^a = 3^{-6}$$

$$3^{2a} = 3^{-6}$$

Since the bases on both sides are the same, they will cancel.

$$2a = -6$$

Divide both sides by 2

$$\frac{2a}{2} = \frac{-6}{2}$$

$$a = -3$$

Question 63:   *Solve the equation* $10^y = 0.0001$

<u>Solution</u>

$$10^y = 0.0001$$

Note: $0.0001 = 10^{-4}$

$$10^y = 10^{-4}$$

Since the bases on both sides are the same, they will cancel.

$$y = -4$$

Question 64:   *Solve the equation* $32^x = 0.25$

<u>Solution</u>

$$32^x = 0.25$$

0.25 to fraction will be equal to $\frac{25}{100} = \frac{1}{4} = 4^{-1}$

Find the lowest common multiple for the bases

$$32 = 2 \times 2 \times 2 \times 2 \times 2 = 2^5$$

$$4 = 2 \times 2 = 2^2$$

$$32^x = 4^{-1}$$

Therefore;

$$(2^5)^x = (2^2)^{-1}$$

$$2^{5x} = 2^{-2}$$

Since the bases on both sides are the same, they will cancel.

$$5x = -2$$

Divide through by 5

$$\frac{5x}{5} = \frac{-2}{5}$$

$$x = -2$$

Question 65:   *Solve the equation $3^{-x} = 243$*

Solution
Find the lowest common multiple of the base on both sides

$$243 = 3 \times 3 \times 3 \times 3 \times 3 = 3^5$$
$$3^{-x} = 3^5$$

Since both sides have the same base, the bases will cancel out.

Therefore;

$$-x = 5$$
$$x = -5$$

Question 66: **Solve for x in the equation below.$25^{5x} = 62$**

Solution
From the equation given. We are to solve for x,

To begin with, we find the lowest common multiple for the bases on both sides.

$$25 = 5 \times 5 = 5^2$$
$$625 = 5 \times 5 \times 5 \times 5 = 5^4$$

$$25^{5x} = 625$$
$$(5^2)^{5x} = 5^4$$
$$5^{2 \times 5x} = 5^4$$
$$5^{10x} = 5^4$$

Question 67: If $\frac{4}{2^x} = 64^x$, find the value x.

Solution

$$\frac{4}{2^X} = \frac{64^X}{1} \ (cross \ mutiply) \ s$$

$$4 = 64^x \times 2^x$$

Find the lowest common base for both R.H.S and L.H.S bases.

$$4 = 2 \times 2 = 2^2$$

$$64 = 2 \times 2 \times 2 \times 2 \times 2 \times 2 = 2^6$$

$$2 = 2^1$$

$$4 = 64^x \times 2^x$$

$$2^2 = (2^6)^x \times 2^x$$

$$2^2 = 2^{6x} \times 2^x$$

$$2^2 = 2^{6x+x}$$

$$2^2 = 2^{7x}$$

$$\frac{2}{7} = \frac{7x}{7}$$

$$x = \frac{2}{7}$$

Question 68:
Solve for the value of x in the equation, $10^x = \frac{1}{0.001}$

Solution

$$10^x = \frac{1}{10^{-3}}$$

$$10^x = 10^{-(-3)}$$

$$10^x = 10^3$$

$$10^x = 10^3$$

$$x = 3$$

Question 69: if $(25)^{x-1} = 64(\frac{5}{2})^6$, find the value of x.

Solution

$$(25)^{x-1} = 64(5/2)^6$$

$$25^{x-1} = 64\left(56/26\right).$$

Find the lowest common base for those bases

$$25 = 5 \times 5 = 5^2$$

$$64 = 2 \times 2 \times 2 \times 2 \times 2 \times 2 = 2^6$$

$$(5^2)^{x-1} = 2^6 \times 5^6/2^6$$

$$5^{2(x-1)} = 5^6$$

$$5^{2x-2} = 5^6$$

$$2x - 2 = 6$$

$$2x = 6 + 2$$

$$2x = 8$$

$$\frac{2x}{2} = \frac{8}{2}$$

$$x = 4$$

Question 70:
Solve for "a" in the equation $10^a \times 5^{(2a-2)} \times 4^{(a-1)} = 1$

Solution

$$10^a \times 5^{(2a-2)} \times 4^{(a-1)} = 1$$

$$(2 \times 5)^a \times 5^{(2a-2)} \times 2^{2(a-1)} = 1$$

$$2^a \times 5^a \times 5^{(2a-2)} \times 2^{2a-2} = 1$$

Combine like terms

$$2^a \times 2^{2a-2} \times 5^a \times 5^{(2a-2)} = 1$$

$$2^{a+(2a-2)} \times 5^{a+(2a-2)} = 1$$

$$2^{a+2a-2} \times 5^{a+2a-2} = 1$$

$$2^{3a-2} \times 5^{3a-2} = 1$$

$$(2 \times 5)^{3a-2} = 1$$

$$(10)^{3a-2} = 1$$

Note :

$$10^0 = 1$$

$$(10)^{3a-2} = 10^0$$

$$10^{3a-2} = 10^0$$

Since both sides have a common base, the bases will cancel out.

$$3a - 2 = 0$$

$$3a = 0 + 2$$

$$3a = 2$$

Divide both sides by 3

$$\frac{3a}{3} = \frac{2}{3}$$

$$a = \frac{2}{3}$$

Question 71: If $\dfrac{9^{2x-1}}{27^{x+1}} = 1$, find the value of x.

Solution

$$\frac{9^{2x-1}}{27^{x+1}} = 1$$

Find the lowest common multiple for the bases

$$9 = 3 \times 3 = 3^2$$

$$27 = 3 \times 3 \times 3 = 3^3$$

$$\frac{9^{2x-1}}{27^{x+1}} = 1$$

Cross multiply

$$(3^2)^{2x-1} = (3^3)^{x+1}$$

$$3^{4x-2} = 3^{3x+3}$$

Since the bases on both sides are the same, the bases will cancel out.

Therefore, we have;

$$4x - 2 = 3x + 3$$

$$4x - 3x = 3 + 2$$

$$x = 5$$

Question 72: *Solve for x in the equation* $8x^{-2} = \frac{2}{25}$

Solution

$$8x^{-2} = \frac{2}{25}$$

Divide through by 8

$$\frac{8x^{-2}}{8} = \frac{2}{25} \div 8$$

$$x^{-2} = \frac{2}{25} \times \frac{1}{8} = \frac{2}{200} = \frac{1}{100}$$

$$x^{-2} = \frac{1}{100} = \frac{1}{10^2} = 10^{-2}$$

$$x^{-2} = 10^{-2}$$

Since the indexes are the same, they will cancel out.

Therefore;

$$x = 10$$

Question 73: If $27^{x+2} \div 9^{x+1} = 3^{2x}$, find the value of x.

<u>Solution</u>

$$27^{x+2} \div 9^{x+1} = 3^{2x}$$

$$(3^3)^{x+2} \div (3^2)^{x+1} = 3^{2x}$$

$$3^{3x+6} \div 3^{2x+2} = 3^{2x}$$

$$3^{3x+6-(2x+2)} = 3^{2x}$$

$$3^{3x+6-2x-2} = 3^{2x}$$

$$3^{3x-2x+6-2} = 3^{2x}$$

$$3^{x+5} = 3^{2x}$$

Since the bases on both sides are the same, they will cancel out.

$$x + 5 = 2x$$

$$5 = 2x - x$$

$$5 = x$$

$$x = 5$$

Question 74: Find the value of a, in $2^{3a+1} = 1$

Solution

$$2^{3a+1} = 1$$

From the law of indices

$$2^0 = 1$$

Substitute 2^0 for 1 in the equation above.

$$2^{3a+1} = 2^0$$

Since the bases on both sides are the same, the bases will cancel out.

$$3a + 1 = 0$$

$$3a = 0 - 1$$

$$3a = -1$$

Divide through by 3

$$\frac{3a}{3} = \frac{-1}{3}$$

$$a = \frac{-1}{3}$$

Question 75:
Find the value of y in the equation $5^{y-1} = 0.2$

Solution

Convert 0.2 into a fraction

$$\frac{2}{10} = \frac{1}{5} = 5^{-1}$$

$$5^{y-1} = 0.2$$
$$5^{y-1} = 5^{-1}$$

Since the bases on both sides are the same, they will cancel out

$$y - 1 = -1$$
$$y = -1 + 1$$

$$y = 0$$

Question 76: Solve for t. $4^t = 2^{3t+3}$

Solution

$$4^t = 2^{3t+3}$$

The lowest common multiple of

$$4 = 2 \times 2 = 2^2$$

Therefore,

$$2^{2(t)} = 2^{3t+3}$$
$$2^{2t} = 2^{3t+3}$$

Since the bases on both sides are the same, they will cancel out.

$$2t = 3t + 3$$
$$2t - 3t = 3$$
$$-t = 3$$

Multiply through by " – "

$$t = -3$$

EXERCISE 2

Solve the following equations.

1. $9^x = 273$

2. $3x = 81x^{\frac{-1}{2}}$, find the value of x.

3. $2^{a-1} = 16,$ *solve for a.*

4. $y^{-2} = 9$

5. $4^{\frac{x}{2}} = 64^{\frac{2}{3}}$, solve for x.

6. $\frac{1}{4}(16^n) = 64$; find the value of n

7. $36^{x-1} = 216^{x+1}$; *hence, find the value of x*

8. $(27)^{\frac{a}{2}} = 3^{\frac{3}{8}} \times 9^{\frac{3}{4}}$; *solve for a.*

9. $5^{y-1} = 625; solve\ for\ y$

10. *Solve for x in* $8^{-1} \times 2^{2x+1} = 64$

Chapter Four

EXPONENTIAL EQUATION PROBLEMS

Some exponential equations problems are reduced to a quadratic form $ax^2 + bx + c$, where a, b, and c are coefficients of x. x is a factor that is to be found for the equation to be balanced.

Miscellaneous Questions and solutions

Question 77: *Solve the following exponential equation*

$$5^{2x} - 26(5^x) + 25 = 0$$

Solution

To solve the question, the exponential equation is reduced to quadratic form.

From the question, let 5^x be replaced with **y**

$5^{2x} - 26(5^x) + 25 = 0$

$(5^x)^2 - 26(5^x) + 25 = 0$

$y^2 - 26y + 25 = 0$

Resolving a quadratic equation, you have to look for two factors.

Solve the quadratic equation

$$y^2 - 26y + 25 = 0$$

Find two factors that, when summed, gives **-26**, and when multiplied, it gives the product of the coefficient of **y² (1)** and **25**.

Sum	product
-25-1=-26	-25*-1=+25

The two factors are -25 and -1, substitute -26y for

$$-25y - 1y \text{ in the equation.}$$

$$y^2 - 25y - y + 25 = 0$$

$$y(y - 25) - 1(y - 25) = 0$$

$$(y - 1)(y - 25) = 0$$

$$y - 1 = 0 \text{ or } y - 25 = 0$$

$$y = 1 \text{ or } 25$$

$$\text{Recall that;}$$

$$y = 5^x$$

$$\text{Therefore;}$$

$5^x = 1$	or	$5^x = 25$
$5^x = 5^0$		$5^x = 5^2$

Since the bases on both sides are the same, they will cancel.

$$x = 0 \text{ or } 2$$

Question 78: **If it is given that $5^{x+1} + 5^x = 150$ then the value of x will be?**

<u>Solution</u>

$$5^{x+1} + 5^x = 150$$

$$5^x \cdot 5^1 + 5^x = 150$$

$$let\ y = 5^x$$

$$y \cdot 5^1 + y = 150$$

$$5y + y = 150$$

$$\frac{6y}{6} = \frac{150}{6}$$

$$y = 25$$

Recall

$$y = 5^x$$

$$5^x = 25$$

$$5^x = 5^2$$

$$\mathbf{x = 2}$$

Question 79: *If $9^{(x-\frac{1}{2})} = 3^{x^2}$, Find the value of x.*

<u>**Solution**</u>

$$9^{(x-\frac{1}{2})} = 3^{x^2}$$

$$(3^2)^{(x-\frac{1}{2})} = 3^{x^2}$$

$$3^{2(x-\frac{1}{2})} = 3^{x^2}$$

$$3^{2x-1} = 3^{x^2}$$

Since the bases are the same, they will cancel out.

Therefore, we have;

$$2x - 1 = x^2$$

$$x^2 - 2x + 1 = 0$$

Solving the quadratic equation.

Find two factors that, when **summed**, gives **-2** (the coefficient of **x**), and when **multiplied**, it gives the product of the coefficient of x² (1) and 1, which equals **1**.

Sum = -2   **product =+1**

-1-1=-2   -1*(-1) = +1

The two factors are -1 and -1, replace -2x with -x-x in the equation.

$$x^2 - 2x + 1 = 0$$
$$x^2 - x - x + 1 = 0$$
$$x(x - 1) - 1(x - 1) = 0$$
$$(x - 1)(x - 1) = 0$$
$$x - 1 = 0$$
$$x = 0 + 1 = 1$$

Therefore

$$x = 1.$$

Question 80:   Solve the following exponential equation

$$(3^x)^2 + 2(3^x) - 3 = 0$$

Solution

$$(3^x)^2 + 2(3^x) - 3 = 0$$

From the question, let 3ˣbe replaced with y

$$(3^x)^2 + 2(3^x) - 3 = 0$$
$$y^2 + 2y - 3 = 0$$

Resolving a quadratic equation, you have to look for two factors.

Solve the quadratic equation

$$y^2 + 2y - 3 = 0$$

Find two factors that, when **summed**, gives +2 (the coefficient of **y**), and when **multiplied**, it gives the product of the coefficient of y^2 (1) and -3, which equals **-3**.

Sum = +2   **product =-3**

+3-1=+2   +3*(-1) = -3

The two factors are +3 and -1, replace -2y with +3y-1y in the equation.

$$y^2 + 3y - y - 3 = 0$$
$$y(y + 3) - 1(y + 3) = 0$$
$$(y - 1)(y + 3)$$
$$y - 1 = 0 \; or \; y + 3 = 0$$
$$y = 1 \; or - 3$$

Recall that;

$$y = 3^x$$

Therefore;

$$3^x = 1 \qquad or \qquad 3^x = -3$$
$$3^x = 3^0 \qquad \qquad \text{no solution}$$

Since the bases on both sides are the same, they will cancel.

$$x = 0$$

Question 81: *Solve the following exponential equation*

$$2^{2x} + 2^{x+1} - 8 = 0$$

Solution

$$2^{2x} + 2^{x+1} - 8 = 0$$

$$(2^x)^2 + (2^x).2^1 - 8 = 0$$

$$(2^x)^2 + 2(2^x) - 8 = 0$$

From the equation above, let 2^x be replaced with y

$$(2^x)^2 + 2(2^x) - 8 = 0$$

$$y^2 + 2y - 8 = 0$$

Resolving a quadratic equation, you have to look for two factors.

Solve the quadratic equation

$$y^2 + 2y - 8 = 0$$

Find two factors that, when summed, gives -2 (the coefficient of y) and when multiplied, it gives the product of the coefficient of y^2 (1) and -8, which equals -8.

Sum = +2	**product= -8**
+4-2=+2	+4*(-2) =-8

The two factors are +4 and -2, substitute +2y for +4y-2y in the equation.

$$y^2 + 4y - 2y - 8 = 0$$

$$y(y + 4) - 2(y + 4) = 0$$

$$(y - 2)(y + 4) = 0$$

$$y - 2 = 0 \ or \ y + 4 = 0$$

$$y = 2 \ or \ y = -4$$

$$y = 2 \; or - 4$$

Recall that;

$$y = 2^x$$

Therefore;

$2^x = 2$	or	$2^x = -4$
$2^x = 2^1$		no solution

Since the bases on both sides are the same, they will cancel.

$$x = 1$$

Question 82: **Solve the following exponential equation**

$$2^{2x} - 5(2^{x+1}) + 16 = 0$$

Solution

$$2^{2x} - 5(2^{x+1}) + 16 = 0$$

$$(2^x)^2 - 5(2^x). 2^1 + 16 = 0$$

$$(2^x)^2 - (5 \times 2)(2^x) + 16 = 0$$

$$(2^x)^2 - 10(2^x) + 16 = 0$$

From the equation above, let 2^x be replaced with **y**

That is; $y = 2^x$

$$(2^x)^2 - 10(2^x) + 16 = 0$$

$$y^2 - 10y + 16 = 0$$

Solve the quadratic equation

$$y^2 - 10y + 16 = 0$$

Find two factors that, when summed, gives -10 (the coefficient of **y**) and when multiplied, it gives the product of the coefficient of y² (1) and 16, which equals **16**.

Sum = -10	product= 16
-8-2=-10	-8*(-2) =+16

The two factors are **-8** and **-2**, substitute **-10y** for **-8y-2y** in the equation.

$$y^2 - 8y - 2y + 16 = 0$$
$$y(y - 8) - 2(y - 8) = 0$$
$$(y - 2)(y - 8) = 0$$
$$y - 2 = 0 \ or \ y - 8 = 0$$
$$y = 2 \ or \ y = 8$$
$$y = 2 \ or \ 8$$

Recall that;

$$y = 2^x$$

Therefore;

$2^x = 2$	or	$2^x = 8$
$2^x = 2^1$		$2^x = 2^3$

Since the bases on both sides are the same, they will cancel.

$x = 1$	or	$x = 3$
		$x = 1 \ or \ 3$

Question 83: **Solve the following exponential equation**

$$2^{2x+1} - (33)2^x + 16 = 0$$

Solution

$$2^{2x+1} - (33)2^x + 16 = 0$$

$$2^1 . 2^{2x} - (2^x).33 + 16 = 0$$

$$2(2^{2x}) - (2^x).33 + 16 = 0$$

From the equation above, let 2^x be replaced with **y**

$$2(2^{2x}) - (2^x).33 + 16 = 0$$

$$2(2^x)^2 - (2^x).33 + 16 = 0$$

$$2y^2 - 33y + 16 = 0$$

Resolving a quadratic equation, you have to look for two factors.

Solve the quadratic equation

$$2y^2 - 33y + 16 = 0$$

Find two factors that, when summed, gives **-33**y and when multiplied, it gives the product of the coefficient of y^2 **(2)** and **16.**

Sum=-33	**product=+32**
-32-1=-33	-32*(-1) =+32

The two factors are -32 and -1, substitute -33y for -32y-1y in the equation.

$$2y^2 - 32y - y + 16 = 0$$

$$2y(y - 16) - 1(y - 16) = 0$$

$$(2y - 1)(y - 16) = 0$$

$2y - 1 = 0 \ or \ y - 16 = 0$

$2y = 1 \ or \ y = 16$

$y = \frac{1}{2} \ or \ 16$

Recall that;

$y = 2^x$

Therefore;

$2^x = \frac{1}{2}$ or $2^x = 16$

$2^x = 2^{-1}$ $2^x = 2^4$

Since the bases on both sides are the same, they will cancel.

$x = -1$ or $x = 4$

$x = -1 \ or \ 4$

Question 84:   **Solve the following exponential equation**

$$2^{2x} - (5)2^x + 4 = 0$$

Solution

$$2^{2x} - 5(2^x) + 4 = 0$$

$$2^{2x} - (2^x).5 + 4 = 0$$

$$(2^x)^2 - 5(2^x) + 4 = 0$$

From the equation above, let 2^x be replaced with y

$$(2^x)^2 - 5(2^x) + 4 = 0$$

$$y^2 - 5y + 4 = 0$$

Solve the quadratic equation

$$y^2 - 5y + 4 = 0$$

Find two factors that, when summed, gives -5y and when multiplied, it gives the product of the coefficient of y² (1) and 4.

Sum=-5   product=+4

-4-1=-5y   -4*(-1) =+4

The two factors are -4 and -1, substitute -5y for

-4y-1y in the equation.

$$y^2 - 4y - y + 4 = 0$$

$$y(y - 4) - 1(y - 4) = 0$$

$$(y - 1)(y - 4) = 0$$

$$y - 1 = 0 \; or \; y - 4 = 0$$

$$y = 1 \; or \; y = 4$$

$$y = 1 \; or \; 16$$

Recall that;

$$y = 2^x$$

Therefore;

$2^x = 1$	or	$2^x = 4$
$2^x = 2^0$		$2^x = 2^2$

Since the bases on both sides are the same, they will cancel.

$$x = 0 \quad \text{or} \quad x = 2$$

$$x = 0 \; or \; 2$$

Question 85: *Solve the following exponential equation*

$$2^{2x+1} - (15)2^x = 8$$

Solution

$$2^{2x+1} - (15)2^x = 8$$

$$2^{2x}.2^1 - 15(2^x) - 8 = 0$$

$$2(2^x)^2 - 15(2^x) - 8 = 0$$

From the equation above, let 2^x be replaced with **y**

$$2(2^x)^2 - 15(2^x) - 8 = 0$$

$$2y^2 - 15y - 8 = 0$$

Solve the quadratic equation

$$2y^2 - 15y - 8 = 0$$

Find two factors that, when summed, gives -15y and when multiplied, it gives the product of the coefficient of y^2 (2) and - 8.

Sum=-15   **product = -16**

-16y+1y=-15y   -16*(+1) =-16

The two factors are **-16** and **+1**, substitute **-15y** for

-16y+1y in the equation.

$$2y^2 - 16y + y - 8 = 0$$

$$2y(y - 8) + 1(y - 8) = 0$$

$$(2y + 1)(y - 8) = 0$$

$$2y + 1 = 0 \ or \ y - 8 = 0$$

$$2y = -1 \ or \ y = 8$$

$$y = \frac{-1}{2} \ or \ 8$$

Recall that;

$$y = 2^x$$

Therefore;

$$2^x = \frac{-1}{2} \qquad or \qquad 2^x = 8$$

$$2^x = -2^{-1} \qquad\qquad 2^x = 2^3$$

No solution

Since the bases on both sides are the same, they will cancel.

$$x = 3$$

Question 86: *Solve the following exponential equation*

$$3^{2x+1} - 28(3^{x-1}) + 1 = 0$$

Solution

$$3^{2x+1} - 28(3^{x-1}) + 1 = 0$$

$$(3^{2x})3^1 - 28(3^x \div 3^1) + 1 = 0$$

$$3(3^x)^2 - 28\left(\frac{3^X}{3^1}\right) + 1 = 0$$

Multiply through by 3

$$3 \times 3(3^x)^2 - (3 \times 28)\left(\frac{3^X}{3^1}\right) + (3 \times 1) = (0 \times 3)$$

$$9(3^x)^2 - 28(3^x) + 3 = 0$$

From the question, let 3^x be replaced with y

$$9(3^x)^2 - 28(3^x) + 3 = 0$$

$$9y^2 - 28y + 3 = 0$$

Solve the quadratic equation

$$9y^2 - 28y + 3 = 0$$

Find two factors that, when summed, gives -28 (the coefficient of **y**) and when multiplied, it gives the product of the coefficient of y^2 (9) and 3, which equals **+27**.

Sum = -28	product =+27
-27-1=-28	-27*(-1) = +27

The two factors are **-27** and **-1**, substitute **-28y** for

-27y-1y in the equation.

$$9y^2 - 27y - y + 3 = 0$$

$$9y(y - 3) - 1(y - 3) = 0$$

$$(9y - 1)(y - 3)$$

$$9y - 1 = 0 \text{ or } y - 3 = 0$$

$$9y = 1 \text{ or } 3$$

$$y = \frac{1}{9} \text{ or } 3$$

Recall that;

$$y = 3^x$$

Therefore;

$$3^x = \frac{1}{9} \qquad \text{or} \qquad 3^x = 3$$

$$3^x = \frac{1}{3^2} \qquad\qquad 3^x = 3^1$$

$$3^x = 3^{-2} \qquad\qquad 3^x = 3^1$$

Since the bases on both sides are the same, they will cancel.

$$x = -2 \qquad\qquad x=1$$

$$x = -2 \text{ or } 1$$

Question 87: Solve the following exponential equation

$$3^{2x-1} - 28(3^{x-2}) + 1 = 0$$

Solution

$$3^{2x-1} - 28(3^{x-2}) + 1 = 0$$

$$(3^{2x}) \div 3^1 - 28(3^x \div 3^2) + 1 = 0$$

$$\frac{(3^x)^2}{3^1} - 28\left(\frac{3^X}{3^2}\right) + 1 = 0$$

$$\frac{(3^x)^2}{3} - 28\left(\frac{3^X}{9}\right) + 1 = 0$$

Multiply through by 9

$$9 \times \frac{(3^x)^2}{3} - (9 \times 28)\left(\frac{3^X}{9}\right) + (9 \times 1) = (0 \times 9)$$

$$3(3^x)^2 - 28(3^x) + 9 = 0$$

From the question, let 3^x be replaced with y

That is; $y = 3^x$

$$3(3^x)^2 - 28(3^x) + 9 = 0$$

$$3y^2 - 28y + 9 = 0$$

Solve the quadratic equation

$$3y^2 - 28y + 9 = 0$$

Find two factors that, when summed, gives -28 (the coefficient of **y**) and when multiplied, it gives the product of the coefficient of y² (3) and 9, which equals **+27**.

Sum = -28	**product =+27**
-27-1=-28	-27*(-1) = +27

The two factors are **-27** and **-1**, substitute **-28y** for

-27y-1y in the equation.

$$3y^2 - 27y - y + 9 = 0$$

$$3y(y - 9) - 1(y - 9) = 0$$

$$(3y - 1)(y - 9)$$

$$3y - 1 = 0 \; or \; y - 9 = 0$$

$$3y = 1 \text{ or } 9$$

$$y = \frac{1}{3} \text{ or } 9$$

Recall that;

$$y = 3^x$$

Therefore;

$$3^x = \frac{1}{3} \qquad \text{or} \qquad 3^x = 9$$

$$3^x = \frac{1}{3} \qquad \qquad \qquad 3^x = 3^2$$

$$3^x = 3^{-1} \qquad \qquad \qquad 3^x = 3^2$$

Since the bases on both sides are the same, they will cancel.

$$x = -1 \qquad \qquad \qquad x=2$$

$$x = -1 \text{ or } 2$$

Question 88: Solve the following exponential equation

$$3^{2x-3} - 4(3^{x-2}) + 1 = 0$$

Solution

$$3^{2x-3} - 4(3^{x-2}) + 1 = 0$$

$$(3^{2x}) \div 3^3 - 4(3^x \div 3^2) + 1 = 0$$

$$\frac{(3^x)^2}{3^3} - 4\left(\frac{3^x}{3^2}\right) + 1 = 0$$

$$\frac{(3^x)^2}{27} - 4\left(\frac{3^x}{9}\right) + 1 = 0$$

Multiply through by 27

$$27 \times \frac{(3^x)^2}{27} - (27 \times 4)\left(\frac{3^x}{9}\right) + (27 \times 1) = (0 \times 27)$$

$$(3^x)^2 - 12(3^x) + 27 = 0$$

From the question, let 3^x be replaced with y

That is; $y = 3^x$

$$(3^x)^2 - 12(3^x) + 27 = 0$$

$$y^2 - 12y + 27 = 0$$

Solve the quadratic equation

$$y^2 - 12y + 27 = 0$$

*Find two factors that, when summed, gives -12 (the coefficient of **y**) and when multiplied, it gives the product of the coefficient of y² (1) and 27, which equals **+27**.*

Sum = -12	product =+27
-9-3=-12	-9*(-3) = +27

*The two factors are **-9** and **-3**, substitute **-12y** for*

-9y-3y *in the equation.*

$$y^2 - 9y - 3y + 27 = 0$$

$$y(y - 9) - 3(y - 9) = 0$$

$$(y - 3)(y - 9)$$

$$y - 3 = 0 \text{ or } y - 9 = 0$$

$$y = 3 \text{ or } 9$$

$$y = 3 \text{ or } 9$$

Recall that;

$$y = 3^x$$

Therefore;

$$3^x = 3 \qquad \text{or} \qquad 3^x = 9$$

$$3^x = 3 \qquad\qquad\qquad 3^x = 3^2$$

$$3^x = 3^1 \qquad\qquad\qquad 3^x = 3^2$$

Since the bases on both sides are the same, they will cancel.

$$x = 1 \qquad\qquad\qquad x=2$$

$x = 1$ **or 2**

Question 89: *Solve the following exponential equation*

$$3^{2x+3} - (3^{x+2}) - 3^{x+1} + 1 = 0$$

Solution

$$3^{2x+3} - (3^{x+2}) - 3^{x+1} + 1 = 0$$

$$(3^{2x}) \times 3^3 - (3^x \times 3^2) - 3^x \times 3^1 + 1 = 0$$

$$27(3^{2x}) - 3^2(3^x) - 3^1(3^x) + 1$$

From the question, let 3^x be replaced with y

That is; $y = 3^x$

$$27(3^x)^2 - 9(3^x) - 3(3^x) + 1 = 0$$

$$27(3^x)^2 - 12(3^x) + 1 = 0$$

$$27y^2 - 12y + 1 = 0$$

Solve the quadratic equation

$$27y^2 - 12y + 1 = 0$$

Find two factors that, when summed, gives -12 (the coefficient of **y**), and when multiplied, it gives the product of the coefficient of y² (27) and 1, which equals **+27**.

Sum = -12   product =+27

-9-3=-12   -9*(-3) = +27

The two factors are -9 and -3, substitute -12y for

-9y-3y in the equation.

$$27y^2 - 9y - 3y + 1 = 0$$

$$9y(3y - 1) - 1(3y - 1) = 0$$

$$(9y - 1)(3y - 1)$$

$$9y - 1 = 0 \; or \; 3y - 1 = 0$$

$$9y = 1 \; or \; 3y = 1$$

$$y = \frac{1}{9} \; or \; y = \frac{1}{3}$$

Recall that;

$$y = 3^x$$

Therefore;

$$3^x = \frac{1}{9} \quad\quad or \quad\quad 3^x = \frac{1}{3}$$

$$3^x = \frac{1}{3^2} \quad\quad\quad\quad\quad 3^x = 3^{-1}$$

$$3^x = 3^{-2} \quad\quad\quad\quad\quad\quad 3^x = 3^{-1}$$

Since the bases on both sides are the same, they will cancel.

$$x = -1 \quad\quad\quad\quad\quad x=-2$$

$$x = -1 \; or \; -2$$

Question 90:	*Solve the following exponential equation*

$$3^x + 3^{1-x} = 4$$

<u>**Solution**</u>

$$3^x + 3^{1-x} = 4$$

$$3^x + (3^1 \div 3^x) = 4$$

$$3^x + \frac{3^1}{3^x} = 4$$

$$3^x + \frac{3}{3^x} = 4$$

Multiply through by 3^x

$$(3^x \times 3^x) + (\frac{3}{3^x} \times 3^x) = (4 \times 3^x)$$

$$(3^x)^2 + 3 = 4(3^x)$$

$$(3^x)^2 - 4(3^x) + 3 = 0$$

From the question, let 3^x be replaced with y

That is; $y = 3^x$

$$(3^x)^2 - 4(3^x) + 3 = 0$$

$$(3^x)^2 - 4(3^x) + 3 = 0$$

$$y^2 - 4y + 3 = 0$$

Solve the quadratic equation

$$y^2 - 4y + 3 = 0$$

*Find two factors that, when summed, gives -4 (the coefficient of **y**), and when multiplied, it gives the product of the coefficient of y^2 (1) and 3, which equals **+3**.*

Sum = -4   **product =+3**

-3-1=-4   -3*(-1) = +3

The two factors are -3 and -1, replace -4y with

-3y-1y in the equation.

$$y^2 - 3y - y + 3 = 0$$

$y(y-3) - 1(y-3) = 0$

$(y-1)(y-3)$

$y - 1 = 0 \ or \ y - 3 = 0$

$y = 1 \ or \ y = 3$

$y = 1 \ or \ y = 3$

Recall that;

$y = 3^x$

Therefore;

$3^x = 1$ or $3^x = 3$

$3^x = 3^0$ $3^x = 3^1$

$3^x = 3^0$ $3^x = 3^1$

Since the bases on both sides are the same, they will cancel.

$x = 0$ $x=1$

$x = 0 \ or \ 1$

Question 91:   *Solve the following exponential equation*

$$2^{2x+3} + 1 = (9)2^x$$

Solution

$2^{2x+3} + 1 = (9)2^x$

$2^{2x}.2^3 + 1 = (9)2^x$

$8(2^x)^2 + 1 = (9)2^x$

$8(2^x)^2 - (9)2^x + 1 = 0$

From the question, let 2^x be replaced with **y**

That is; $y = 2^x$

$$8(2^x)^2 - (9)2^x + 1 = 0$$
$$8y^2 - 9y + 1 = 0$$

Solve the quadratic equation

$$8y^2 - 9y + 1 = 0$$

Find two factors that, when **summed**, gives -9 (the coefficient of **y**), and when **multiplied**, it gives the product of the coefficient of y² (8) and 1, which equals **+8**.

Sum = -9	**product =+8**
-8-1=-9	-8*(-8) = +8

The two factors are **-8** and **-1**, replace -9y with
-8y-1y in the equation.

$$8y^2 - 9y + 1 = 0$$
$$8y^2 - 8y - y + 1 = 0$$
$$8y(y - 1) - 1(y - 1) = 0$$
$$(8y - 1)(y - 1) = 0$$
$$8y - 1 = 0 \text{ or } y - 1 = 0$$
$$8y = 1 \text{ or } y = 1$$
$$y = \frac{1}{8} \text{ or } 1$$

Recall that;

$$y = 2^x$$

Therefore;

$$2^x = \frac{1}{8} \qquad\qquad \text{or} \qquad 2^x = 1$$

$$2^x = \frac{1}{2^3} \qquad\qquad\qquad 2^x = 2^0$$

$$2^x = 2^{-3} \qquad\qquad\qquad 2^x = 2^0$$

Since the bases on both sides are the same, they will cancel.

$$x = -3 \qquad \text{or} \qquad\qquad x = 0$$

$$x = -3 \text{ or } 0$$

Question 92   Solve the following exponential equation

$$2^{2x} - (3)2^x + 2 = 0$$

Solution

$$2^{2x} - (3)2^x + 2 = 0$$

$$(2^x)^2 - (3)2^x + 2 = 0$$

From the question, let 2^x be replaced with y

That is; $y = 2^x$

$$(2^x)^2 - (3)2^x + 2 = 0$$

$$y^2 - 3y + 2 = 0$$

Solve the quadratic equation

$$y^2 - 3y + 2 = 0$$

Find two factors that, when **summed**, gives -3 (the coefficient of **y**), and when **multiplied**, it gives the product of the coefficient of y² (1) and 2, which equals **+2**.

Sum = -3   **product =+2**

-2-1=-3   -2*(-1) = +2

The two factors are -2 and -1, replace -3y with -2y-1y in the equation.

$$y^2 - 3y + 2 = 0$$
$$y^2 - 2y - y + 2 = 0$$
$$y(y - 2) - 1(y - 2) = 0$$
$$(y - 1)(y - 2) = 0$$
$$y - 1 = 0 \ or \ y - 2 = 0$$
$$y = 1 \ or \ y = 2$$
$$y = 1 \ or \ 2$$

Recall that;

$$y = 2^x$$

Therefore;

$2^x = 1$	or	$2^x = 2$
$2^x = 2^0$		$2^x = 2^1$
$2^x = 2^0$		$2^x = 2^1$

Since the bases on both sides are the same, they will cancel.

$x = 0$	or	$x = 1$
		$x = 0 \ or \ 1$

Question 93: Solve the following exponential equation

$$4(4^x - 2^x) + 1 = 2^x$$

Solution

$$4(2^{2x} - 2^x) + 1 = 2^x$$
$$4((2^x)^2 - 2^x)) + 1 = 2^x$$

From the question, let 2^x be replaced with y

That is; $y = 2^x$

$$4((2^x)^2 - 2^x)) + 1 = 2^x$$
$$4(y^2 - y) + 1 = y$$
$$4y^2 - 4y + 1 = y$$
$$4y^2 - 4y - y + 1 = 0$$
$$4y^2 - 5y + 1 = y$$

Solve the quadratic equation

$$4y^2 - 5y + 1 = 0$$

*Find two factors that, when **summed**, gives -5 (the coefficient of **y**), and when **multiplied**, it gives the product of the coefficient of y² (4) and 1, which equals **+4**.*

Sum = -5   **product =+4**

-4-1=-3   -4*(-1) = +2

*The two factors are **-4** and **-1**, replace -5y with*

-4y-1y *in the equation.*

$$y^2 - 5y + 1 = 0$$
$$y^2 - 4y - y + 1 = 0$$
$$y(y - 4) - 1(y - 1) = 0$$
$$(y - 1)(y - 4) = 0$$
$$y - 1 = 0 \; or \; y - 4 = 0$$
$$y = 1 \; or \; y = 4$$
$$y = 1 \; or \; 4$$

Recall that;

$$y = 2^x$$

Therefore;

$$2^x = 1 \qquad \text{or} \qquad 2^x = 4$$

$$2^x = 2^0 \qquad\qquad 2^x = 2^2$$

$$2^x = 2^0 \qquad\qquad 2^x = 2^2$$

Since the bases on both sides are the same, they will cancel.

$$x = 0 \qquad \text{or} \qquad\qquad x = 2$$

$$x = 0 \text{ or } 2$$

Question 94: **Solve the following exponential equation**

$$2^{x^2-2} = 16(2^{5x})$$

Solution

$$2^{x^2-2} = 16(2^{5x})$$

The lowest common multiple of base 16

$$16 = 2 \times 2 \times 2 \times 2 = 2^4$$

Replace 16 by 2^4 in the equation

$$2^{x^2-2} = 2^4(2^{5x})$$

$$2^{x^2-2} = 2^4 \times (2^{5x})$$

$$2^{x^2-2} = 2^{4+5x}$$

Since both sides have the same base, the bases will cancel out.

Therefore;

$$x^2 - 2 = 4 + 5x$$

$$x^2 - 2 - 4 - 5x = 0$$

$$x^2 - 5x - 6 = 0$$

The equation is quadratic

$$x^2 - 5x - 6 = 0$$

Find two factors that, when **summed**, gives -5 (the coefficient of **x**), and when **multiplied**, it gives the product of the coefficient of x² (1) and -6, which equals **-6**.

Sum = -5   **product =-6**

-6+1=-5   -6*(1) = -6

The two factors are **-6** and **1**, replace **-5x** with **-6x-1x** in the equation.

$$x^2 - 5x - 6 = 0$$
$$x^2 - 6x + x - 6 = 0$$
$$x(x - 6) + 1(x - 6) = 0$$
$$(x + 1)(x - 6) = 0$$
$$x + 1 = 0 \ or \ x - 6 = 0$$
$$x = -1 \ or \ x = 6$$
$$x = -1 \ or \ 6$$
$$x = -1 \ or \ 6$$

Problem 95: *Solve the following exponential equation*
$$2^{2x-3} - 2^{x-2} - 1 = 0$$

Solution

$$2^{2x-3} - (2^{x-2}) - 1 = 0$$

$$(2^{2x}) \div 2^3 - (2^x \div 2^2) - 1 = 0$$

$$\frac{(2^x)^2}{2^3} - \left(\frac{2^x}{2^2}\right) - 1 = 0$$

$$\frac{(2^x)^2}{8} - \left(\frac{2^x}{4}\right) - 1 = 0$$

Multiply through by 8

$$8 \times \frac{(2^x)^2}{8} - (8)\left(\frac{2^x}{4}\right) - (8 \times 1) = (0 \times 8)$$

$$(2^x)^2 - 2(2^x) - 8 = 0$$

From the question, let 2^x be replaced with y

That is; $y = 2^x$

$$(2^x)^2 - 2(2^x) - 8 = 0$$

$$y^2 - 2y - 8 = 0$$

Solve the quadratic equation

$$y^2 - 2y - 8 = 0$$

*Find two factors that, when summed, gives -2 (the coefficient of **y**) and when multiplied, it gives the product of the coefficient of y^2 (1) and -8, which equals **-8**.*

Sum = -2   product =-8

+4-2=-2   +4*(-2) = -8

The two factors are **+4** and **-2**, substitute **-2y** for **+4y-2y** in the equation.

$$y^2 + 4y - 2y - 8 = 0$$
$$y(y + 4) - 2(y + 4) = 0$$
$$(y - 2)(y + 4)$$
$$y - 2 = 0 \ or \ y + 4 = 0$$
$$y = 2 \ or -4$$

Recall that;

$$y = 2^x$$

Therefore;

$$2^x = 2 \qquad or \qquad 2^x = -4$$
$$2^x = 2^1 \qquad\qquad no \ solution$$

Since the bases on both sides are the same, they will cancel.

$$x = 1$$

$x = 1$

Question 96: *Solve the following exponential equation*

$$2^x + 2^{-x} = 2$$

Solution

$$2^x + 2^{-x} = 2$$

$$2^x + \frac{1}{2^x} = 2$$

Multiply through by 2^x

$$2^x \times 2^x + 2^x \times \frac{1}{2^x} = 2 \times 2^x$$

$$2^{2x} + 1 = 2(2^x)$$

$$(2^x)^2 + 1 = 2(2^x)$$

$$(2^x)^2 + 2(2^x) + 1 = 0$$

From the question, let 2^x be replaced with y

That is; $y = 2^x$

$$(2^x)^2 + (2)2^x + 1 = 0$$

$$y^2 + 2y + 1 = 0$$

Solve the quadratic equation

$$y^2 + 2y + 1 = 0$$

Find two factors that, when **summed**, gives +2 (the coefficient of **y**), and when **multiplied**, it gives the product of the coefficient of y² (1) and 1, which equals **+1**.

Sum = +2   **product =+**

+1+1=2   1*(1) = +1

The two factors are **+1** and **+1**, replace **2y** with

+1y+1y in the equation.

$$y^2 + y + y + 1 = 0$$
$$y^2 + y + y + 1 = 0$$
$$y(y + 1) + 1(y + 1) = 0$$
$$(y + 1)(y + 1) = 0$$
$$y + 1 = 0 \ or \ y + 1 = 0$$
$$y = -1 \ or \ y = -1$$
$$y = -1 \ or - 1$$

Recall that;

$$y = 2^x$$

Therefore;

$$2^x = -1 \qquad \qquad or \qquad 2^x = -1$$

No solution   no solution

Question 97:   Solve for the value of x, in the equation

$$3^{x^2+2} = 27^x$$

Solution

$$3^{x^2+2} = 27^x$$

If $27 = 3 \times 3 \times 3 = 3^3$ replace 27^x with 3^3

$$3^{x^2+2} = (3^3)^x$$
$$3^{x^2+2} = 3^{3x}$$

Since the bases on both sides are the same, they will cancel out.

$$x^2 + 2 = 3x$$
$$x^2 - 3x + 2 = 0$$

The equation is quadratic

$$x^2 - 3x + 2 = 0$$

Find two factors that, when **summed**, gives -3 (the coefficient of **x**), and when **multiplied**, it gives the product of the coefficient of x² (1) and +2, which equals **+2**.

Sum = -3   **product =+2**

-2-1=-3   -2*(-1) = +2

The two factors **are -2** and **-1**, replace -3x with **-2x-1x** in the equation.

$$x^2 - 3x + 2 = 0$$
$$x^2 - 2x - x + 2 = 0$$
$$x(x - 2) - 1(x - 2) = 0$$
$$(x - 1)(x - 2) = 0$$
$$x - 1 = 0 \ or \ x - 2 = 0$$
$$x = 1 \ or \ x = 2$$
$$x = 1 \ or \ 2$$

$$x = 1 \ or \ 2$$

Question 98: *Solve the following exponential equation*

$$5^{2x} + 1 = 26(5^{x-1})$$

Solution

$$5^{2x} + 1 = 26(5^{x-1})$$

$$5^{2x} - 26(5^{x-1}) + 1 = 0$$

$$5^{2x} - 26(5^x \div 5^1) + 1 = 0$$

$$(5^x)^2 - 26\left(\frac{5^x}{5^1}\right) + 1 = 0$$

Multiply through by 5

$$5 \times (3^x)^2 - (5 \times 26)\left(\frac{5^x}{5}\right) + (5 \times 1) = (0 \times 5)$$

$$5(5^x)^2 - 26(5^x) + 5 = 0$$

From the question, let 3ˣbe replaced with y

$$5(5^x)^2 - 26(5^x) + 5 = 0$$

$$5y^2 - 26y + 5 = 0$$

Solve the quadratic equation

$$5y^2 - 26y + 5 = 0$$

*Find two factors that, when summed, gives -26 (the coefficient of **y**) and when multiplied, it gives the product of the coefficient of y² (5) and 5, which equals **+25**.*

Sum = -26	product =+25
-25-1=-26	-25*(-1) = +25

The two factors **are -25** and **-1**, substitute +25 for

-25y-1y in the equation.

$$5y^2 - 25y - y + 5 = 0$$
$$5y(y - 5) - 1(y - 5) = 0$$
$$(5y - 1)(y - 5)$$
$$5y - 1 = 0 \; or \; y - 5 = 0$$
$$5y = 1 \; or \; 5$$
$$y = \frac{1}{5} \; or \; 5$$

Recall that;

$$y = 5^x$$

Therefore;

$$5^x = \frac{1}{5} \qquad or \qquad 5^x = 5$$
$$5^x = \frac{1}{5^1} \qquad\qquad 5^x = 5^1$$
$$5^x = 5^{-1} \qquad\qquad 5^x = 5^1$$

Since the bases on both sides are the same, they will cancel.

$$x = -1 \qquad\qquad x=1$$

$x = -1$ or 1

Question 99:   **Solve the system of equation**

$$8^{4-x} = 2^{1-y}$$

$$3^{x+1} = 9^{-y}$$

Solution

$$8^{4-x} = 2^{1-y} \qquad \ldots\ldots\ldots.(1)$$

$$3^{x+1} = 9^{-y} \qquad \ldots\ldots\ldots.(2)$$

Find the lowest common multiple for the bases in both equations.

$$8 = 2 \times 2 \times 2 = 2^3$$

$$9 = 3 \times 3 = 3^2$$

From equation 1 and 2, we have

$$2^{3(4-x)} = 2^{1-y} \qquad \ldots\ldots\ldots.(3)$$

$$3^{x+1} = 3^{2(-y)} \qquad \ldots\ldots\ldots.(4)$$

$$2^{12-3x} = 2^{1-y} \qquad \ldots\ldots\ldots(5)$$

$$3^{x+1} = 3^{-2y} \qquad \ldots\ldots\ldots.(6)$$

Since the bases on either side of both equations 5 and 6 are the same, the bases will cancel.

$$12 - 3x = 1 - y \qquad \ldots\ldots\ldots.(7)$$

$$x + 1 = -2y \qquad \ldots\ldots\ldots\ldots.(8)$$

Therefore, we can solve the simultaneous equations to find the value of x and y.

In equation 8,

$$x = -2y - 1.$$

Substitute $-2y - 1$ for x in **equation 7**,

Therefore;

$$12 - (-2y - 1) = 1 - y$$

$$12 + 2y + 1 = 1 - y$$

$$12 + 2y - y + 1 - 1 = 0$$

$$12 + y = 0$$

$$y = -12$$

Substitute -12 for y, in **equation (8)**

$$x + 1 = -2y$$

$$x = -2y - 1$$

$$x = -2(-12) - 1$$

$$x = 24 - 1$$

$$x = 23.$$

Therefore;

$$x = 23, y = -12$$

Problem 100:   Solve the system of equation

$$2^{x+y} = 32$$

$$3^{3y-x} = 27$$

Solution

$$2^{x+y} = 32 \qquad \text{............ (1)}$$

$$3^{3y-x} = 27 \qquad \text{............ (2)}$$

Find the lowest common multiple for the bases in both equations.

$$32 = 2 \times 2 \times 2 \times 2 \times 2 = 2^5$$

$$27 = 3 \times 3 \times 3 = 3^3$$

From equation 1 and 2, we have

$$2^{x+y} = 2^5 \rightarrow (3)$$

$$3^{3y-x} = 3^3 \rightarrow (4)$$

Since the bases on either sides of both equations 3 and 4 are the same, the bases will cancel.

$$x + y = 5 \rightarrow (5)$$

$$3y - x = 3 \rightarrow (6)$$

Therefore, we can solve the simultaneous equations to find the value of x and y.

In equation 5,

$$x = 5 - y.$$

Substitute $5 - y$ for x in equation 6,

Therefore;

$$3y - (5 - y) = 3$$

$$3y - 5 + y = 3$$

$$3y + y - 5 = 3$$

$$3y + y = 3 + 5$$

$$4y = 8$$

Divide both sides by 4

$$\frac{4y}{4} = \frac{8}{4}$$

$$y = 2.$$

Substitute 2 for y, in equation (5)

$$x + 2 = 5$$

$x = 5 - 2$

$x = 3.$

Therefore; $x = 3, y = 2$

EXERCISE 3

Solve the following equations.

1. *solve for x in*; $2^{2x} - 6(2^x) + 8 = 0$

2. *solve for x in;* $2^{2x+1} - 5(2^x) + 2 = 0$

3. $3^{2x} - 4(3^{x+1}) + 27 = 0;\ solve\ for\ x$

A N S W E R S

Exercise 1

1. 2401

2. $35x^4$

3. $4x^3$

4. 1

5. $\frac{4}{3}$

6. 84

7. $\frac{13b}{a}$

8. 0.5

9. $2a^{-3}$

10. $\frac{64}{27}$

Exercise 2

1. $\frac{5}{2}$

2. 9

3. 5

4. $\frac{1}{3}$

5. 2

6. 4

7. $\frac{5}{2}$

8. $\frac{5}{4}$

9. 5.

10. 4

Exercise 3

1. $x = 1 \ or \ 2$

2. $x = -1 \ or \ 1$

3. $x = 1 \ or \ 2$

About the Author

Samuel Adegboye is a dedicated scientist and lecturer who has devoted a significant portion of his life to academia and practical applications. He has actively assisted individuals of all ages in resolving scientific and mathematical problems. With a background in Electrical/Electronic Engineering, Samuel successfully tackled various science subjects, including Mathematics, Physics, Chemistry, and more, during his academic journey. As the visionary behind Kunlektra Academy, his primary objective is to empower young individuals academically, regardless of their backgrounds. Samuel firmly holds the belief that no subject or topic is inherently challenging; rather, he emphasizes the importance of understanding the fundamentals, as it helps prevent numerous errors along the way.

Acknowledgments

First of all, my appreciation goes to God, Almighty, for the opportunity to collate this manuscript. This book was published, with the support of God Almighty and some persons he used to be part of the success of this book. I am grateful for some friends, colleagues, and co-members in encouraging and supporting me to start the work, persevere with it, and finally to publish it.

.

THANKS FOR PATRONAGE

Improve your Math Skills with other Books

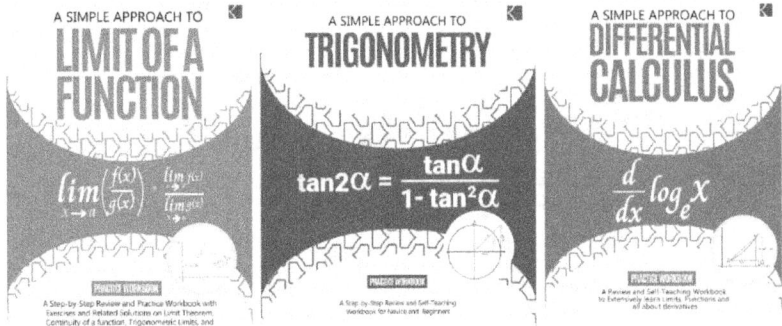

www.ingramcontent.com/pod-product-compliance
Lightning Source LLC
Chambersburg PA
CBHW060332130626
46553CB00003B/979